Postharvest Technology of Tropical Fruits and Vegetables

THE AUTHORS

Dr. Surajit Mitra is a Professor and Former Head of the Department of Postharvest Technology under the Faculty of Horticulture at Bidhan Chandra Krishi Viswavidyalaya (State Agricultural University), West Bengal, India. He acquired his M. Sc. (Ag.) in Horticulture and Ph. D. degree from Bidhan Chandra Krishi Viswavidyalaya in 1993 and 1998 respectively. He was awarded several National and International Fellowships, and participated in International Training Courses on Postharvest Technology at Kasetsart University, Thailand; Postharvest Education Foundation, USA; Agricultural Research Organization, Israel; and Wageningen University, Netherlands. He also attended and presented several research papers, and delivered lead lectures on nutrition and postharvest technology in International Conferences in England, Germany, Belgium, Luxembourg, Sri Lanka and Bangladesh, and both in National and International Conferences in India. Prof. Mitra delivered a keynote speech and chaired a Technical Session in the 17th World Congress on 'Nutrition and Food Chemistry- 2018' at London, UK. He has published 71 research papers in peer reviewed national and international journals, authored 20 book chapters and 4 books on different aspects of postharvest technology of horticultural crops. Prof. Mitra has supervised 11 students of M. Sc. (Hort.) and 9 students of Ph. D. in Postharvest Technology. He has developed two varieties each of sweet potato and eddoe taro, and one variety each of dasheen taro, elephant foot yam and swamp taro. Among various recognitions, Chaudhary Debi Lal Outstanding AICRP Award from Indian Council of Agricultural Research as a team member of AICRP on Tuber Crops; Best Paper Award by International Society for Horticultural Science for a collaborative research work with colleagues in Venezuela; Outstanding Scientist Award-2018 from the Society of Tropical Agriculture, New Delhi, India, Eminent Scientist Award-2016 from Samagra Vikas Welfare Society, Lucknow, Uttar Pradesh, India, Outstanding Scientist Award-2015 from Venus International Foundation, Chennai, India; Achiever Award-2015 from SADHNA, Dr. YS Parmar University of Horticulture and Forestry, Solan, Himachal Pradesh, India are noteworthy. Dr. Mitra is an Editorial Board Member of International Journal of Forestry and Horticulture, Permanent Honorary Board Member of Society for advancement of Human and Nature, and Invited editor for special issues on nutrition and postharvest technology in some Open Access Journals. He is also a Life Member of the Indian Academy of Horticultural Sciences, Indian Science Congress Association, Society for the Advancement of Horticulture, Indian Society for Root Crops, and Crop and Weed Science Society.

Professor T.K. Bose awarded M.Sc. (Ag.), Ph.D. degree from Calcutta University, West Bengal, India. Former Horticulturist and Secretary of the Agri-Horticultural Society of India; Project Coordinator, All India Coordinated Floriculture Improvement Project (ICAR); Professor and Head, Department of Horticulture; Dean, Faculty of Agriculture; Director of Research, Bidhan Chandra Krishi Viswavidyalaya, West Bengal; Member, State Agriculture Commission, West Bengal; Member of Scientific Advisory Committees of ICAR, CSIR (Govt. of India) and Government of West Bengal. Awarded D.Sc. (Honoris Causa) from OUAT, Bhubaneshwar, Odisha and UBKV, Coochbehar, West Bengal, India. Guided around 45 students for their Ph.D. degree, published more than 250 research papers and authored 30 books on horticultural crops.

Postharvest Technology of Tropical Fruits and Vegetables

Surajit Mitra

Editor

M. Sc. (Ag) in Horticulture, Ph. D.
Professor and Former Head
Department of Postharvest Technology
Bidhan Chandra Krishi Viswavidyalaya (Agril. University)
Mohanpur-741252, Nadia, West Bengal, India

T.K. Bose

Editor Emeritus

Former Professor (Horticulture) and Director of Research
Faculty of Horticulture
Bidhan Chandra Krishi Viswavidyalaya (Agril. University)
Mohanpur-741252, Nadia, West Bengal, India

2023

Daya Publishing House®

A Division of

Astral International Pvt. Ltd.

New Delhi – 110002

Cataloging in Publication Data--DK
Courtesy: D.K. Agencies (P) Ltd. <docinfo@dkagencies.com>

Postharvest technology of tropical fruits and vegetables / [editors], Surajit Mitra, T.K. Bose.
pages cm
Contributed articles.
Includes bibliographical references and index.
ISBN 9789359191652 (HB)

1. Tropical fruit--Postharvest technology. 2. Tropical fruit--Postharvest technology--India. 3. Vegetables--Postharvest technology. 4. Vegetables--Postharvest technology--India. I. Mitra, Surajit, editor. II. Bose, T. K., editor.
LCC SB360.P67 2023 | DDC 634.04 23

Published by : **Daya Publishing House®**
A Division of
Astral International Pvt. Ltd.
– ISO 9001:2015 Certified Company –
4736/23, Ansari Road, Darya Ganj
New Delhi-110 002
Ph. 011-43549197, 8130496929
E-mail: info@astralint.com
Website: www.astralint.com

Dedicated to the Farmers and Scientists

Who continue to render their uninterrupted services even during Covid-19 pandemic across the world

Acknowledgements

On my behalf as an editor and author, and also on behalf of other contributors, I express my deep sense of indebtedness to the scientists, authors and publishers for consulting their articles, journals and books during compilation of different chapters of this book.

I owe my deepest gratitude to the Department of Postharvest Technology, Bidhan Chandra Krishi Viswavidyalaya, Mohanpur, West Bengal for providing facilities to conduct research works related to some topics of this book.

I express my sincere thanks to all contributors and the Editor Emeritus Prof. T.K. Bose for constant support and encouragement in publishing this book.

I would like to thank my Ph. D. students who carried out research works under my guidance for their efforts in collecting relevant references for many chapters in this book.

I am thankful to my wife and daughter for their constant encouragement and support during the course of writing and editing the chapters of this book.

I am also thankful to Mr Prateek Mittal Astral International Pvt Limited for publishing this book.

Kalyani

Surajit Mitra

From the Desk of Prof. T.K. Bose

In this book, an effort has been made to include various aspects of postharvest technology with an objective to extend the shelf life of fruits and vegetables, and thereby to reduce postharvest losses. This edition has included up-to-date infotmation on postharvest handling systems with meaningful illustrations. This book is going to be considered as valuable contribution on the subject, and will be useful for the students, researchers, teachers and extension workers.

T.K. Bose

Editor Emeritus

Preface

About forty five per cent of the world's harvested fruits and vegetables is lost due to high perishability on account of high moisture content, and limited application of postharvest technology in developing countries. Therefore, "Postharvest Technology" needs to be considered as an effective means to reduce huge postharvest losses and thereby to enhance food and nutrition security.

In this book, an attempt has been made to cover the most important aspects of postharvest management of tropical fruits and vegetables including new development in postharvest technology along with the conventional technologies.

In this book, an attempt has also been made to highlight major causes of spoilage of fresh and processed products with proper postharvest management for reducing loss and maintaining quality of harvested produce to make fruits and vegetables available all the year round either in fresh or processed form.

Contents of this book are based on syllabi of different courses on postharvest technology offered both at undergraduate and postgraduate levels in the agricultural and horticultural universities. The endeavour in compiling this book will only be successful if the students, researchers, teachers, extension workers and interested people find this book useful.

There is always a scope for improvement of any book. Fruitful suggestion or constructive criticism is invited for further improvement of this book.

Surajit Mitra

Editor

Contents

List of Contributors

Surajit Mitra

Professor and Former Head, Department of Postharvest Technology, Bidhan Chandra Krishi Viswavidyalaya (Agril. University), Mohanpur-741252, Nadia, West Bengal, India

Sonja Veljovic

Food Scientist, Institute of General and Physical Chemistry, Belgrade, P.O. Box 551, 11001 Belgrade, Serbia

Ghan Shyam Abrol

Assistant Professor, College of Horticulture and Forestry, Rani Lakshmi Bai Central Agricultural University, Jhansi-284003, Uttar Pradesh, India

Amit Kumar Singh

Assistant Professor, Postharvest Technology, College of Horticulture, Banda University of Agriculture and Technology, Banda-210001, Uttar Pradesh, India

Arvind Kumar Chaurasiya

Assistant Professor, Faculty department of Horticulture, North-Eastern Hill University, Tura Campus, Meghalaya-794002

Arghya Mani

Department of Postharvest Technology, Bidhan Chandra Krishi Viswavidyalaya, Mohanpur-741252, Nadia, West Bengal, India

Saša Despotovic

Assist. Professor, University of Belgrade, Faculty of Agriculture, Nemanjina 4, 11080 Belgrade, Serbia

Manisha Ch. Momin

Department of Postharvest Technology, Bidhan Chandra Krishi Viswavidyalaya, Mohanpur-741252, Nadia, West Bengal, India

Hugo O. Ramírez-Guerrero

Professor, Zamorano University, Honduras

Nguyen Thanh Tung

Deputy Head, Division of Biotechnology, Physiology and Postharvest Technology, Southern Horticultural Research Institute, Tien Giang Province, Vietnam

Alemmoa. R. Jamir

Department of Plantation, Spices, Medicinal and Aromatic Crops, Bidhan Chandra Krishi Viswavidyalaya, Mohanpur-741252, Nadia, West Bengal, India

J. Suresh Kumar

Scientist, PC Unit, AICRP on Tuber Crops, ICAR-Central Tuber Crops Research Institute, Sreekariyam, Thiruvananthapuram, Kerala-695 017, India

T. Krishnakumar

Scientist, Division of Crop Utilization, ICAR-Central Tuber Crops Research Institute, Sreekariyam, Thiruvananthapuram, Kerala-695 017, India

R. Arutselvan

Scientist, Division of Crop Protection, Regional Centre, ICAR-Central Tuber Crops Research Institute, Dumuduma, Bhubaneswar, Odisha-751019, India

M.S. Sajeev

Principal Scientist and Head, Division of Crop Utilization, ICAR-Central Tuber Crops Research Institute, Sreekariyam, Thiruvananthapuram, Kerala-695 017, India

S. Sunitha

Principal Scientist, PC Unit, AICRP on Tuber Crops, ICAR-Central Tuber Crops Research Institute, Sreekariyam, Thiruvananthapuram, Kerala-695 017, India

M. Preema Devi

Department of Pomology and Postharvest Technology, Uttar Banga Krishi Vishwavidyalaya, Pundibari, Cooch Behar, West Bengal, India

J. Aresh

Department of Pomology and Postharvest Technology, Uttar Banga Krishi Vishwavidyalaya, Pundibari, Cooch Behar, West Bengal, India

Aparna G. S

Department of Postharvest Technology, College of Agriculture, Vellayani, Kerala Agriculture University, Kerala, India

Maria Fernanda Berlingieri Durigan

Scientist, Postharvest and Plant Industrialization, Embrapa Instrumentação, São Carlos, São Paulo State – Brazil

Biswajit Karmakar

Department of Postharvest Technology, Bidhan Chandra Krishi Viswavidyalaya, Mohanpur-741252, Nadia, West Bengal, India

Chapter 1

Fresh Fruit and Vegetable Industry, and Postharvest Losses: An Overview

Introduction

Fresh fruits and vegetables as major source of essential vitamins and minerals constitute important parts of the human diet. As the animal foods do not contain any fibre, we solely depend on plant foods, especially different fruits, vegetables and tubers for dietary fibre, mostly the undigested cellulose which is essential for bowel movement, and some are prebiotic which promote growth of gut bacteria and help in controlling blood sugar and lipids. Desirable dietary intake of nutrients as per the requirement is imperative to maintain good health. Both the macronutrients like carbohydrates, proteins and fats, and micronutrients like vitamins and minerals along with some phytochemicals are required for formulating a balanced diet to lead a healthy and active life. Some of the fruits and vegetables rich in protein, vitamins, minerals and anti-oxidants, and low in carbohydrates, fats and anti-nutrients seem to help in reducing detrimental effects of high fat and high sugar diets to the hippocampus, a major component of the brain. Being protective food, different fruits and vegetables are gaining popularity in almost all over the world as nutrition-security crop to fight oxidative stress, improve general cognition, reduce cholesterol levels and help in maintaining healthy blood pressure levels. The plant foods including fruits and vegetables are only source of dietary fibre, which is essential for bowel movement, growth of gut bacteria and for controlling blood sugar and lipids.

Fruits and vegetables having high β-carotene and anthocyanins are also gaining importance as antioxidant foods to tackle the problem of vitamin A deficiency, improving visual acuity and liver function, and to prevent the obesity, heart disease and diabetes. But high moisture content and continuation of metabolic activities as living commodities even after harvest cause huge postharvest losses of fruits and vegetables. In spite of huge production of about 300 million tonnes of fruits and vegetables in India, there is a deficit of 20-25 per cent in per capita availability due to postharvest losses from farm to fork. Globally, about one-third of all food that is produced each year is still wasted, and approximately, 45 per cent of all fruits and vegetables grown are not consumed, causing loss of billions of dollars mainly due to inefficient value chains in developing countries. In this chapter, an attempt has been made to discuss a brief account of various aspects of postharvest losses and management systems including typical value chains for perishables and measures to address the inevitable problems for prolonging shelf life of fruits and vegetables to reduce the loss of harvested produce and thereby to boost farmers' income for lifting rural economy, and to increase the affordability of fruits and vegetables for alleviating global malnutrition. Extending storage life of fruits and vegetables with the implementation of improved postharvest technologies and proper monitoring will not only help in arresting food waste and loss, but will be helpful in preventing wastage of natural resources used by the harvested produce, and also in reversing global warming by reducing greenhouse gas emissions due to decomposition of lost and wasted food.

Dietary Allowance should be based on the energy requirement of an individual in terms of the level of energy intake from a food that balances energy expenditure, and bioavailability of a given nutrient from a diet. Inclusion of nutrient-dense fruits and vegetables in regular diet is required for human wellbeing. The attributes of quality of fruits and vegetables like attractive skin colour, excellent flavour, taste, texture, and nutritional properties are associated with consumer acceptance. Maintenance of the quality of harvested fruits and vegetables is very challenging because of their high perishability and short shelf life due to high moisture content, soft texture, high physiological activities and susceptibility to many postharvest diseases and pests. Postharvest deterioration of horticultural produce including fruits and vegetables starts immediately after harvest because of the continuation of physiological processes like transpiration and respiration at the cost of stored water and food. Unlike the plants in the field, compensation of water lost by transpiration, and food degraded by respiration is not possible in the harvested produce. Rapid loss of water after harvest is one of the most serious postharvest problems in maintaining freshness of fruits and vegetables. Huge quantities of the harvested fruits and vegetables are lost due to high perishability on account of high moisture content, and limited application of postharvest technology in developing countries.

India is the world's second largest producer of fruits and vegetables next only to China with a total production of about 300 million tonnes. Yet agriculture, which makes up over 50 per cent of the workforce, generates just 15 per cent of GDP. In India, improper postharvest handling, inadequate cold supply chain, lack of storage facilities and insufficient food processing units lead to 20-40 per cent losses of harvested fruits and vegetables causing a monetary loss of about Rs. 2 lakh Crores every year. Although, as per APEDA (2021), food processing sector is one of the largest sectors in India in terms of production, growth, consumption, and export with a total value of all exported processed food at about Rs. 37,000 Crores (USD 5000 Millions) in 2020-21, fruit and vegetable processing industry is still in rudimentary stage with an estimated use of 7-9 million tonnes of fruits and vegetables comprising only 2.5-3.0 per cent of the total production. In spite of the second largest producer of fruits and vegetables in the world, India is far behind from the countries like Malaysia (83 per cent), Philippines (78 per cent), Brazil (70 per cent), USA (65 per cent), Thailand (30 per cent) and China (23 per cent) in value addition and shelf life extension of fruits and vegetables. So, there is a huge scope for expansion of food processing industry in India.

To feed the ever-increasing population in the world, any preventable loss of food is not acceptable, and should be emphasised for reducing postharvest losses to the possible extent. Along with securing food for all, food safety also needs to be focused as it is directly related with human health. Some of these problems can be solved by improvements in farming practices and appropriate postharvest handling and processing. Although many techniques have been developed, no method is best suitable for all the commodities due to wide variations in nature and texture of fruits and vegetables. Coordinated activities by growers, handlers, storage operators, processors, wholesalers and retailers are required to maintain quality and reduce postharvest loss of perishable commodities. The extent of coordination may vary from very simple in local food supply to complex in global supply chains. Due to multiple uncertainties along the supply chain, microbial contamination may occur leading to spoilage at any stage in the chain from farm to consumer. Therefore, proper harvesting, handling, postharvest treatments, management of storage temperature, relative humidity and gaseous composition are essential to minimize physiological activities, microbial spoilage and to reduce the risk of foodborne illness. Since the trade in fruits and vegetables are gaining momentum in both developed and developing countries, proper preparation of fresh commodities before marketing, and other technological interventions are required to facilitate and stimulate the development of the marketing of fruits and vegetables.

Considering the importance of extension of shelf life of fruits and vegetables for reducing huge postharvest losses, proper measures must be taken against major causative factors like mechanical damage, physiological deterioration and microbial spoilage of fruits and vegetables with proper postharvest handling systems for reducing loss and maintaining quality of harvested produce to make them available all the year round either in fresh or processed form, enhance food and nutrition security, and grasp the opportunity with both hands for economic development. Along with the use of conventional postharvest technologies, prioritization of cold supply chains, nanotechnology, smart packaging, improved storage techniques, non-thermal innovative processing technologies including cold plasma technology, pulsed electric fields, ultrasonication and high pressure processing can be helpful in microbial decontamination without affecting nutritional and sensory characteristics of foods, and in prolonging storage life of fruits and vegetables for reducing the loss of harvested produce, which in turn will boost farmers' income, lift rural economy, and increase the affordability of fruits and vegetables for alleviating global malnutrition. Gene editing to extend shelf-life without loss of quality is also being tried as a promising new avenue for reducing postharvest loss of nutrient-dense fruits and vegetables. Promotion and proper implementation of new technologies for extending storage life of fruits and vegetables will not only help in reducing postharvest losses, but will also be helpful in preventing wastage of natural resources used by the harvested produce. In addition, harmful effect of greenhouse gas in contributing to global warming can be reduced by retarding its emissions due to decomposition of lost and wasted food. Furthermore, not only the increasing population but also the loss of harvested produce cause further demand for food production. If food loss is significantly reduced, the extra pressure for producing more food to meet the increasing demand can be reduced remarkably.

Extent of Postharvest Loss

It is estimated that globally the food loss and waste in horticulture sector accounts for 40 to 50 per cent of the total production, and some 2 billion people, about 30 per cent of the planet's population suffer from micronutrient deficiencies or other effects of inadequate diet (FAO, 2014). In India alone, 30 to 40 per cent of fruit and vegetable grown (40 MMT value to US$ 13 billion) gets wasted in 2005-06 due to the gaps in the cold chain such as inadequate cold storage capacity, poor infrastructure, unavailability of cold storages close to farms, poor logistics infrastructure, *etc.* (Maheshwar and Chanakwa, 2006). These losses increased over the years with the increase in production of fruits and vegetables. The data published by the Associated Chambers of Commerce and Industry of India (ASSOCHAM), based on production and wholesale market price in India, it was found that postharvest losses of fruits and vegetables reached an amount of $33,745 million in 2011-12. Among the main producing states, West Bengal suffered the highest loss of $ 2,163 million followed by Gujarat $ 1,805 million, Bihar 1,702 million and Uttar Pradesh $ 1,633 million (ASSOCHAM, 2013). In general, the food losses are higher in horticultural crops, illustrated by the postharvest losses of 52 per cent in cauliflower, 46 per cent in tomato, 20 per cent in mandarin, 29 per cent in banana and 38 per cent in mango which are produced in the SAARC nations (FAO, 2015). A NASSCOM study of 2019 has suggested that Indian farmers face postharvest losses amounting to a whopping Rs 93,000 crore. Institution of Mechanical Engineers (IME) had conducted a study, "A Tank of Cold: Cleantech Leapfrog to a more food secure world" on farm losses in the country and trying to sentitise cost-effective technologies in cold chain, the most crucial element in stemming farm losses.

India is the world's second largest producer of fruits and vegetables next only to China with a total production of about 300 million tonnes in 2019-2020. Yet agriculture, which makes up over 55 per cent of the workforce, generates just 15-16 per cent of GDP. In India, improper postharvest handling, inadequate cold supply chain, lack of cold storage facilities and insufficient food processing units lead to 20-40 per cent losses of harvested fruits and vegetables causing a monetary loss of about Rs. 2 lakh Crores every year. Although, food processing sector is one of the largest sectors in India in terms of production, growth, consumption, and export with a total value of all exported processed foods at about Rs. 37,000 Crores (USD 5000 Millions) including Rs. 11,375 Crores from processed fruit and vegetable products in 2020-21 (APEDA, 2021), fruit and vegetable processing industry is still in rudimentary stage with an estimated use of 7-9 million tonnes of fruits and vegetables comprising only 2.5-3.0 per cent of the total production. In spite of the second largest producer of fruits and vegetables in the world, India is far behind from the countries like Malaysia (83 per cent), Philippines (78 per cent), Brazil (70 per cent), USA (65 per cent), Thailand (30 per cent) and China (23 per cent) in value addition and shelf life extension of fruits and vegetables. In the financial year 2019-20, 8.31 million tonnes of fruits and vegetables were processed in India, which is expected to reach 16.39 million tonnes by 2024, expanding at a CAGR of 14.84 per cent during 2020- 2024 period (Business Wire, 2020). So, there is a huge scope for expansion of cold chains and food processing industry in India for extending shelf life and reducing postharvest losses of fruits and vegetables. If energy expenses Energy expenses make up about 30 per cent of the total expenses for Indian cold storages compared to 10 per cent in the West. High energy expenses not only make it difficult for development of cold chain facilities and setting up of cold storages but also it is unviable and uneconomical. About 20-25 per cent of the losses can be reduced by transporting the freshly harvested fruits and vegetables in refrigerated containers and storing them in cold storages, thus bridging this gap in the cold chain.

Food and nutritional security is a global agenda and can be attained by providing nutritional and balanced diet to everybody. World Health Organization's recommendation is 400 grams of fruit and vegetables per day/ person. Globally, fruit and vegetable production was 306 grams per person in 2000, was reached 390 grams in 2017. World's total vegetables and fruit production in 2018 was about 1 billion 957 million tonnes (1 billion 89 million tonnes of vegetables and 868 million tonnes of fruits). Assuming current world population of about 7.9 billion, now, fruit and vegetable production per person is about 650g/day, but due to 40-50 per cent loss of harvested fruits and vegetables, the availability is much lower than the recommended quantity.

Types of Postharvest Loss

Food Loss and Food Waste

Food losses refer to the decrease in edible food mass available for human consumption throughout the different segments of the supply chain. In addition to quantitative losses, food products can also face a deterioration of quality, leading to a loss of economic and nutritional value. Food waste refers to food losses resulting from decisions to discard food that still has value. Food waste is most often associated with the behaviour of retailers, the food service sectors and consumers, but food waste and losses take place all along food supply chains (FAO, 2012).

Though wasted food refers to the discarded food due to expiry of date or other reasons, especially in developed countries, it is an integral part of total food loss, and the terms food waste and food loss are interchangeably used. Food loss does not only mean the loss of physical foods, rather it accounts for the waste of water, land, fertilizer, chemicals, labour, money, time, energy and other resources involved in the production of the lost food (Mitra, 2021).

Among different kinds of postharvest loss, some important losses are:

1. **Qualitative loss:** Qualitative loss includes loss of caloric and nutritive value, loss of acceptability by consumers, and loss of edibility of the produce which is more difficult to measure than quantitative loss of fresh commodity. Reduction of qualitative loss is a higher priority than quantitative loss in developed countries where consumer dissatisfaction with produce quality results in a greater percentage of postharvest losses.
2. **Quantitative loss:** Quantitative loss refers to the physical loss of a produce that can be measured in terms of percentage or any other units. This loss is a higher priority than qualitative loss in developing countries.
3. **Psychological loss:** This loss is due to non-acceptance of consumers to new kinds or varieties of crops due to a feeling of repugnance towards a particular crop with a desire to avoid. Human aversion or disapproval and religious taboos accentuate this type of losses.

Different Points of Loss of Fruits and Vegetables

1. **Losses at harvest**: Mechanical injuries and pressure damage cause deterioration in quality of harvested produce, which also act as secondary source of infection.

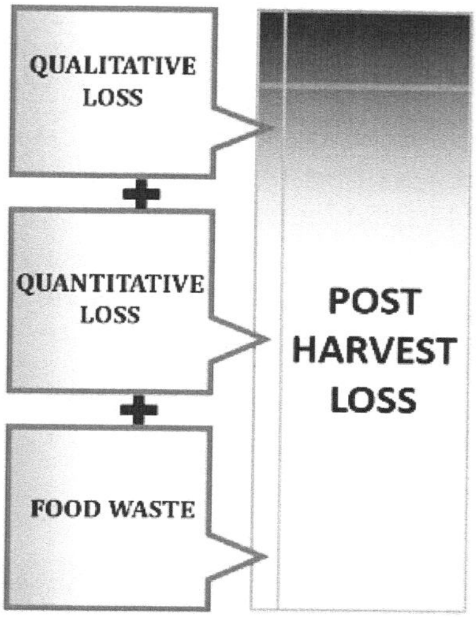

Figure 1. Types of Postharvest Loss.

2. **Losses at the packinghouse**: Damage is caused by chemicals, brushes and wax used during preparation for packing.
3. **Losses during preparation**: Loss is due to peeling of the fruits, vegetables and tubers.
4. **Loss during preservation and processing:** Loss is due to removal of skin, seeds, core and pomace after extraction of juice *etc.*
5. **Losses during storage**: Damage due to chilling injuries, decay, peel disorders, attack of pest and rodents.
6. **Losses during transport:** Damage due to compression, decay, bruising and deformation during loading and unloading.
7. **Losses at market**: Damage is mainly due to wilting, discolouration, over ripening, softening and decay.
8. **Losses at consumers' home**: Damage is due to wilting, over ripening, softening and decay of produce purchased more than the requirement of a family.

Causes of Postharvest Losses

In most of the developing countries, postharvest losses of highly perishable fruits and vegetables due to high moisture content generally ranges from 30- 40 per cent, and in some cases, it goes up to 80 per cent as against 5-20 per cent in developed countries depending upon the crop, cultivar, season and handling systems. Not only producers but also different players or actors involved in postharvest handling systems of horticultural produce must understand that harvested fruits, vegetables and other horticultural crops are living commodities and continue their metabolic activities causing deterioration in quality, which leads to senescence, the ultimate death of cells leaving the produce inconsumable or unusable. Along with the physiological processes, the microbial and environmental factors involved in deterioration also need to be understood clearly. Knowing the causes of postharvest loss of horticultural crops, use of appropriate postharvest techniques can delay this deterioration and extend the shelf-life of harvested produce maintaining desirable qualities. Different causes of postharvest deterioration of harvested produce are shown in Figure 2.

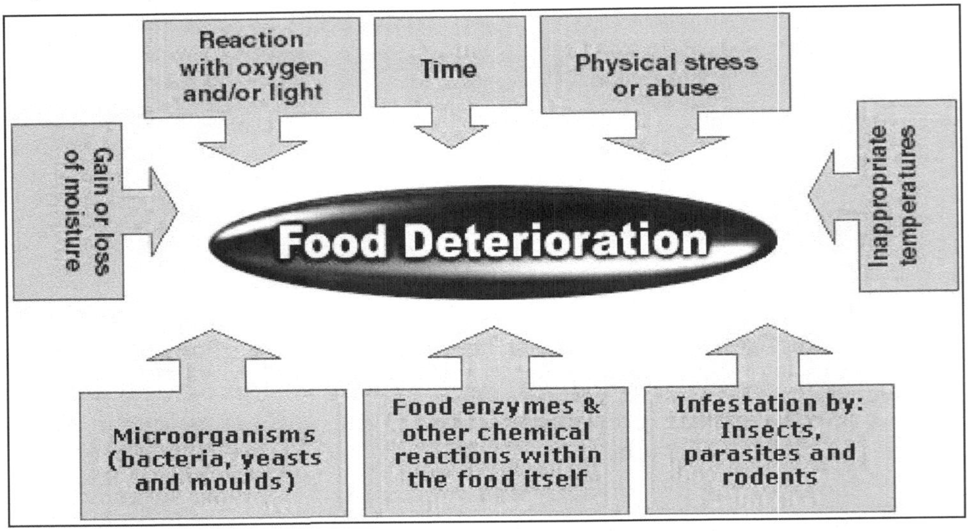

Figure 2: Causes of Postharvest Deterioration of Perishables.

Factors Responsible for Loss of Fruits and Vegetables

1. **Physiological or internal factors:** Causes of deterioration include respiration rate, ethylene production and action, rates of compositional changes affecting colour, texture, flavour, and nutritive value, mechanical injuries, water stress, sprouting and rooting, physiological disorders, and pathological breakdown. The rate of physiological deterioration also depends on several environmental factors.
2. **Microbial or biological factors:** Different microorganisms, mainly bacteria and fungi including moulds and yeasts deteriorate the quality and market value of horticultural crops by causing various postharvest diseases. Apart from microbes, some pests and rodents are also responsible for postharvest losses.
3. **Environmental or external factors:** These causes include temperature, relative humidity, air velocity, atmospheric composition comprising concentrations of oxygen, carbon dioxide and ethylene, and sanitation procedures.

4. **Socioeconomic factors:** Although the internal and external causes of postharvest losses are well understood and many technologies have been developed to reduce these losses, they have not been implemented due to some socioeconomic factors like inadequate facilities for packaging, transportation, storage and marketing.

Apart from the physiological, microbial and environmental causes, some of the important reasons responsible for postharvest losses are as follows-

1. **Lack of Knowledge:** Majority of the growers and handlers are not in the know about the need and merits of using some special tools, equipment and technology in harvesting and postharvest handling of fresh commodity.
2. **Inadequate Packaging:** In most cases, cares are not taken for proper packaging of produces during transportation, storage and marketing due to non-availability of suitable packaging materials or poor economic condition of the growers resulting deterioration in quality of produce.
3. **Inadequate Transportation Facilities:** In most rural areas, roads are not adequate for proper transport of fresh commodities. Transport vehicles and other modes of transportation suitable for perishable crops are also in short supply both for local marketing and export to other countries. Except some marketing organizations and cooperatives, majority of small holding producers do not have their own transport vehicles.
4. **Inadequate Marketing Systems:** Growers produce large quantities of good-quality product but lack of organized markets, commodities fail to reach the consumers with a pace it require causing extensive losses in many locations within developing countries.
5. **Lack of Government Regulations:** Lack of strict governmental control over fraudulent marketing, inadequate regulations covering prop handling procedures and public health aspects during marketing, a no incentives for producing high-quality product or for postharv quality maintenance.
6. **Inadequate Supply of Tools and Equipment:** Needed tools equipment for harvesting and postharvest handling of commodity are not readily available in the domestic markets.
7. **Lack of Information:** Most of the growers and handlers in directly in harvesting, packaging, transporting and marke developing countries have limited or no information regar need and means for maintaining quality.
8. **Poor Maintenance:** In many developing countries, lack of ma and unavailability of spare parts are the major causes functioning of good machineries.

9. **Negligence in Postharvest Research and Development:** Although minimizing postharvest losses of already produced food is more sustainable than increasing production to compensate for these losses, less than 5 per cent of the funding for agricultural research is allocated to postharvest research areas.

Initiatives from both public and private sectors in investing more on infrastructure development especially cold chains, the most crucial element in prolonging shelf life of perishables, and processing units; allocating more funds for postharvest research areas, and emphasising more on capacity building in postharvest management, and food standards for complying with food safety standards can reduce postharvest losses remarkably.

In addition to making infrastructural development, capacity building in food standards for complying with food safety standards can reduce food losses as well as improve health and food security status of a nation. The factors that make food unsafe for consumption like naturally occurring toxins, contaminated water, indiscriminate use of pesticides and other harmful chemicals, and unhygienic harvesting, handling and storage conditions need to be dealt most effectively for maintaining minimum food safety standards and reducing food losses.

In a nutshell, there are several internal and external causes for postharvest deterioration of fruits and vegetables. Some of the damages caused by various factors are easily identified by the consumers, which reduce the market value of the crops. Damages caused by biological factors like pests and diseases, chemical factors like visible external contamination with chemical substance or its toxic and unpleasant flavour, mechanical factors like injures, cuts, bruises and scrapings, physical factors like heating, chilling or freezing and shrivelling, and the physiological factors like sprouting, rooting, senescence, and changes caused by excessive transpiration and respiration are found to reduce the market value as well as shelf-life of harvested fruits and vegetables. Furthermore, some pre-harvest factors such as environmental factors like temperature, relative humidity, rainfall, light, frost *etc.*; genetic factors like type of crop and cultivar; cultural practices like planting time, planting density, fertilization, irrigation and plant protection techniques also affect the postharvest quality and shelf life of a produce. Actually, the combined effect of all these factors decides the rate of deterioration in quality, spoilage, and postharvest performance after harvesting of fruits and vegetables.

As a result, despite the remarkable progress made in increasing food production globally, approximately half of the population in the developing countries does not have access to adequate food supplies due to food losses occurring throughout the supply chain from production, postharvest, processing and marketing. Evidence suggests that these losses tend to be highest in those countries where the need for food is greatest. Stark contrasts exist in managing the available food around the world. In parts of our societies especially in developing countries, there is a shortage of food while in the urban and wealthier communities in developed countries good food is thrown away only because it is beyond its sell-by date. Optimizing the 'farm to fork' chain can considerably reduce the loss of food contributing significantly to food security. Estimates indicate that 30 per cent to 40 per cent of the food produced globally is lost at various stages from production, harvesting, handling, marketing, distribution and processing or wasted because it is never consumed. Without these losses there is sufficient food available to feed the rapidly increasing world population.

Strategies for Extending Storage Life of Fruits and Vegetables

For achieving main objectives of postharvest technology such as maintaining quality of produce, protecting food safety, reducing losses between harvest and consumption, and promoting food and nutrition security, effective management during the postharvest period is more important than the level of sophistication of any given technology. One of the aims of postharvest technology is to reduce the physiological processes for extending shelf-life of harvested produce. Proper control of the external factors that influence the physiological processes by reducing mechanical injury and temperature, maintaining high relative humidity, manipulating atmospheric composition by reducing oxygen level and increasing carbon dioxide, reducing accumulation of ethylene, and imparting physical barrier by waxing or using plastic films can delay the tissue disintegration or senescence phase and extend the shelf-life of fruits and vegetables. Proper control of both the physiological and external factors having direct effect on the ripening, an irreversible process that occur at the cellular level involving a series of anabolic and catalytic reactions including degradation of chlorophyll, production of carotenoids and anthocyanins, conversion of starch to sugars, changes in organic acids, proteins and fats, development of flavours, reduction in tannins and fungistatic compounds, softening of cell wall, and finally collapse or death of cells can retard the rate of deterioration and prolong the storage of fruits and vegetables. The activities in each and every value chain need to be properly monitored for arresting postharvest losses of fruits and vegetables.

Figure 3: Typical Value Chains of Fruits and Vegetables.

Steps of Postharvest System

There are many interacting steps involved in any postharvest system. Harvested commodities are often handled by many people, transported and stored repeatedly between harvest and consumption. While particular practice and the sequence of operations vary for each crop depending on the nature of the crop and intended use, there is a general series of steps in postharvest handling system as shown in Figure 4.

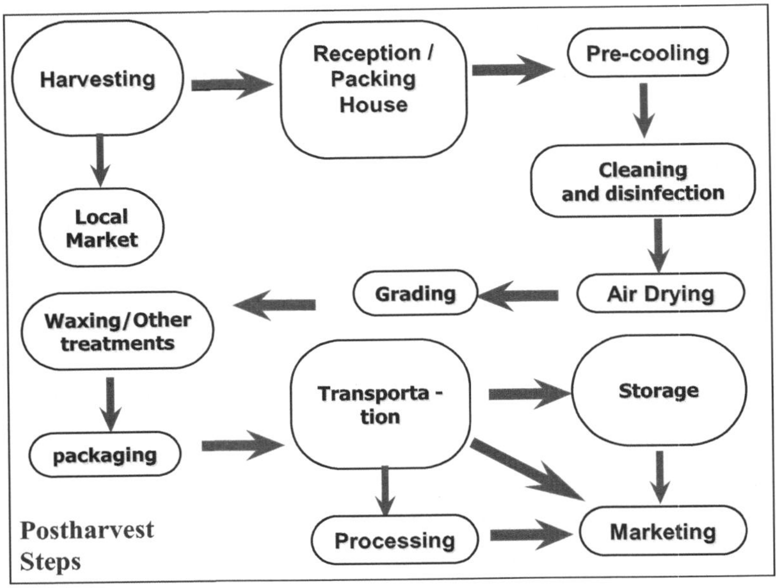

Figure 4: General Steps in Postharvest Handling System (*Source*: Mitra, 2021).

Harvesting

For harvesting a quality produce, care must be taken right from the selection of good genetic materials for planting to proper harvesting of the crop including good agricultural practices (GAP) in-between. If the quality of any crop is not good at the time of harvest, there is no way to improve its quality after harvesting. Postharvest technology only can maintain the quality of harvested fresh produce by reducing the rate of deterioration which starts immediately after harvesting of the crop. Harvesting of fruits and vegetables at proper stage of maturity is very important for its subsequent quality, storage and marketable life. Different maturity indices standardised based on different methods like physical (skin colour, shape, size, firmness and aroma), chemical (sugar, starch and acidity), physiological (rate of respiration), computation (heat unit or degree days), and electronic and other methods (electronic colour sorter, and acoustic and vibration test) can be used for judging proper maturity of fruits and vegetables. During harvesting, care should also be taken to avoid mechanical injury by following suitable methods of harvesting either by hands or machines.

Harvesting should be done at the cooler part of the day. Care should be taken to avoid mechanical injuries. Plastic crates or other suitable containers need to be used to avoid soil contact during harvesting. Harvesting poles are to be used for harvesting from tall trees to avoid dropping. Placing of fruits and vegetables under shade in the field following harvesting till they are transported will help in reducing field heat.

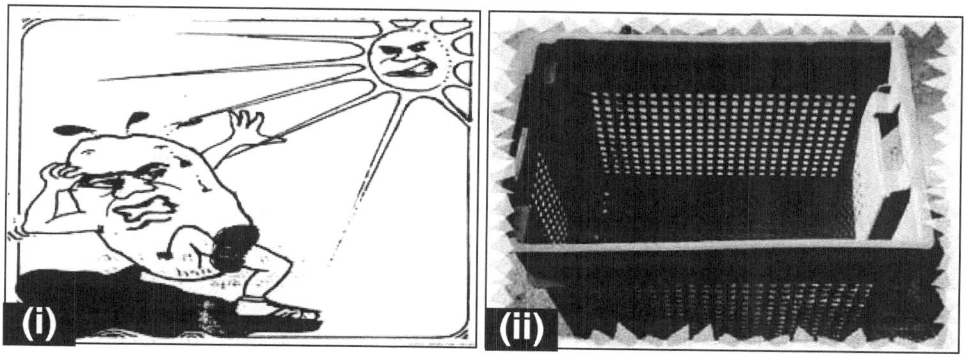

Figure 5: (i) Heat Injured Produce Harvested during Mid-day; (ii) Plastic Crate for Harvesting.

Transportation from Field to Local Market/Packing House

To maintain the quality during transport, the following points must be considered:

1. Proper transportation needs to be arranged before harvesting, if not available on the day of harvesting, the harvested produce will rot on the ground in the field.
2. Adequate transport facilities need to be ensured.
3. Provision for refrigeration in transport system needs to be made, and mix load should be avoided.
4. For local market: Non-refrigerated trucks with ventilation may be allowed, as these losses are relatively minimal because of the short time frame for the local market.
5. For export: Transportation in refrigerated trucks is must.
6. For reducing mechanical injuries during transportation due to compression, plastic crates should be arranged on the platforms.

Figure 6: (i) Improper Transport; (ii) Proper Transport.

Packaging

Proper packaging plays important roles in containment, protection, communication and convenience during movement from one place to another, storage and marketing of fresh commodities as well as processed products. In addition to various functions of traditional packaging, many innovative packaging methods have been developed for extending shelf life of fruits and vegetables.

Packaging is the enclosure of products, items or packages in a wrapped pouch, bag, box, cup, tray, can, tube, bottle or other containers to perform important functions

Commonly Used Packages

1. *Consumer or Retail Package*

i) Flexible Plastic Films- LDPE (Polyethylene), PVC (Poly Vinyl Chloride), PP (Polypropylene) and cellulose acetate films.

ii) Trays with Overwrap- Usually made of LDPE, LLDPE or PVC. The films are semi-permeable and allow exchange of gases for respiration of the product.

iii) Plastic Punnets- Made of PET, PVC or PP

iv) Plastic Net bags (Extruded and Woven)- Generally made of HDPE (High Density Polyethylene) or PA (Polyamide).

v) Foam Sleeve- Plastic tubular film made of polyethylene foam, easily slipped over the individual fruits, provides a cushioning effect.

vi) Light Weight Plastic Crates- Generally made of HDPE or PP. These hygienic, reusable, perforated crates provide ventilation and keep the produce fresh.

vii) Insert trays- Typically formed from molded plastic or pulp to create an insert for a master container that has multiple cavities to contain individual fruit.

viii) Clamshells- Most common polymer used in clamshell containers is polyethylene terephthalate (PET).

2. *Transport/Bulk Packs*

i) Bulk bins- Originally used in harvesting, are now used for shipping and storage of some produce.

ii) Wooden wire bound containers

iii) Corrugated fiberboard cartons

iv) Reusable plastic containers

v) Bamboo Baskets

vi) Jute reinforced plastics (JRP)

Figure 7: (i) Individual Protection of Large Fruits with Plastic Foam Nets (ii) Plastic Punnets; (iii) Plastic Net Bag

Pre-cooling

Pre-cooling is a very important step in the postharvest handling system of perishables for rapid removal of field heat shortly after the harvest. Field heat can be defined as the difference in temperature between the harvested crop and the optimal storage temperature of that produce. The process of pre-cooling is the removal of field heat to arrest the deteriorative processes for maintaining desired quality. Nilanthiand Guleria (2013) observed that precooled fruits exhibited relatively slower loss in weight and decline in fruit firmness as compared to the non-precooled. This may be due to the fact that loss of moisture from pre-cooled commodities is slower if higher relative humidity is maintained in the storage atmosphere. Fruit firmness maintenance in tomatoes treated by pre-cooling was better than that of the tomato fruits which were not precooled (Watkins *et al.*, 1990). This may be because of inhibition or inactivation of cell wall hydrolytic enzymes such as polygalacturonase and pectinesterase or suppression of mRNA synthesis coding for wall softening enzymes in tomatoes treated by pre-cooling. Nunes *et al.* (2005) demonstrated that perishable produce such as berries should be pre-cooled immediately or not more than 2 or 3 hours after harvest to reduce decay and quality loss during storage. Rapid reduction of field temperature and storage near 10°C can delay the ripening and decay. Doshi and Sutar (2010) also reported an additional extension of five days in shelf life of precooled banana stored at 13.5 °C and 80-85 per cent relative humidity. To reduce the field heat, harvested fruits and vegetables should be kept under shade in the field immediately after harvest until they are transported from the field. The results from a study conducted by Kyung *et al.* (2019) demonstrated the effectiveness of pre-cooling to suppress the softening and indicate an impact on microbial community composition and dynamics during storage.

Field heat can cause rapid deterioration of some horticultural crops and therefore, it is desirable to remove this heat as quickly as possible after harvesting.

Influence on Respiration Rate- pre-cooling removes the field heat before storage to reduce the respiration rate and hence deterioration will decline accordingly.

Influence on metabolism- Metabolic rates double for each 10°C rise in temperature.

Effect on ethylene- The reduction in temperature reduces the production and sensitivity of the produce to ethylene.

Points to Remember during Pre-cooling

1. In stacking produce, adequate air-circulation need to be allowed to ensure even cooling.
2. Proper receptacles such as vented boxes and baskets should be used for pre-cooling.
3. Potable water should be used for pre-cooling of fruits and vegetables to minimize any food safety concerns.
4. Ethylene-sensitive produce needs to be separated from ethylene producing one.
5. If a chiller is used for pre-cooling, keeping it closed at all times is required to minimize temperature and relative humidity fluctuations.
6. The pre-cooling facility should be loaded at or below its optimum capacity.

Different Types of Pre-cooling

1. Room cooling with refrigeration
2. Hydro cooling
3. Vacuum cooling
4. Ice cooling
5. Forced-air cooling

Storage

In developing countries, especially in tropical countries, long term storage of fruits and vegetables is a major problem where sufficient cold stores are not available. Along with the microbial spoilage of fruits and vegetables, sprouting in tubers and bulbous crops is a major problem during storage in conventional low cost storage structures. Therefore, proper storage facilities with the provisions for modification of temperature, relative humidity, gaseous concentration and pressure as per the nature of commodity need to be created for prolonging storage of fruits and vegetables. Storage is an important operation to reduce the loss of harvested produce. Owing to high moisture content and metabolic activities they are more liable to spoilage. Only about 2.5-3 per cent of harvested fruits and vegetables are processed into value added products. Storage of fresh horticultural produce is one of the critical problems in many developing countries. Storage helps in orderly marketing avoiding seasonal glut and increases profit to the growers and traders. Proper storage of the perishable fresh produce using appropriate methods can prolong their availability by -

1. Control of respiration rate
2. Control of transpiration
3. Control of sprouting and rooting
4. Control of ripening and senescence
5. Control of microbial spoilage

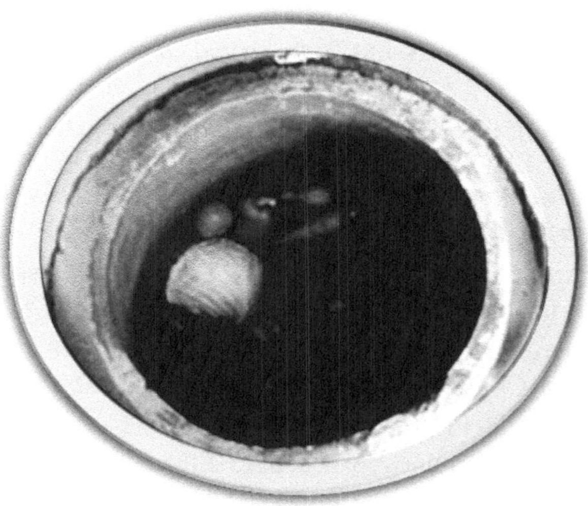

Figure 8: Clay Pot Evaporative Cooler.

Along with the conventional storage structures like cellar storage, natural ventilation type storage, forced air ventilation type storage and storage using evaporative coolers, several improved storage structures developed with the manipulation of temperature, humidity, gaseous concentration and atmospheric pressure can be used for shelf life extension of fruits and vegetables.

References

APEDA. 2021. India export of agro food products. Product group report-processed fruits and vegetables; other processed foods. *Agricultural and Processed Food Products Export Development Authority*, India.

ASSOCHAM (2013) The Associated Chambers of Commerce and Industry of India, *www.assocham.org*.

Business Wire. 2020. Outlook on the fruit and vegetable processing industry in India to 2024- Featuring Dabur India, ITC and Freshtrop Fruits among others - *ResearchAndMarkets.com*, April 24, 2020, Dublin.

Doshi, J. S. and Sutar, R. F. 2010. Studies on pre-cooling and storage banana for extension of shelf life. *J. Agric. Eng.*, **47** (2): 14-19.

FAO. 2012. The role of producer organizations in reducing food loss and waste. *International Year of Cooperatives, Issue Brief Series*, Rome.

FAO. 2014. Better nutrition, better lives. *Second International Conference on Nutrition*, 19-21 November 2014, FAO Headquarters, Rome, Italy

FAO. 2015. FAO. 2015. Reduction of postharvest losses in horticultural chains in SAARC Countries. *Regional Technical Cooperation Programme* (*TCP*) *Project Report*, Rome, Italy.

Fellows, P. J. 2018. *Food Processing Technology: Principles and Practices*. Elsevier. Pp. 279-280.

Kyung, M. P., Hyo, J. K., Sang, S. K., Sang, B. L., Mooncheol, J., Kee, J. P. and Minseon, K. 2019. Effect of temperature treatment on postharvest quality of the cherry tomato (*Lycopersicon esculentum* var. *cerasiforme*). *Korean J. Food Preserv.*, 26 (6): 595-605

Maheshwar, C. and Chanakwa, T. S. 2006. Postharvest losses due to gaps in cold chain in India – A solution. *Acta Hortic.*, **712:** 777-784

Mitra, S. 2021. Postharvest technology-An overview. **In**: *Postharvest Management of Horticultural Crops*. Surajit Mitra (Ed.), Jaya Publishing House, New Delhi-110089, India, pp. 1-16.

Nilanthi, A. W. R. M. and Guleria, S. P. S. 2013. Effect of pre-cooling, fruit coating and packaging on postharvest quality of apple. *J Food Sci Technol*, 50: 325-331

Nunes, M. C. N., Morais, A. M. M. B., Brecht, J. K., Sargent, S. A. and Bartz, J. A. 2005. Prompt cooling reduces incidence and severity of decay caused by *Botrytis cinerea* and *Rhizopus stolonifer* in strawberry. *Hort Technology*, 15: 153-156

Watkins C. B., Picton S. and Grierson, D. 1990. Stimulation and inhibition of expression of ripening-related mRNAs in tomatoes as influenced by chilling temperatures. *J. Plant Physiol,* 136: 318-323.

Xuetong, F. 2018. *Food Irradiation research and Technology*. Wiley-Blackwell. ISBN 978-0-8138-0209-1.

Chapter 2

New Development in Postharvest Technology

Introduction

Due to multiple uncertainties along the supply chain, microbial contamination may occur leading to spoilage at any stage in the chain from farm to consumer. Therefore, proper harvesting, handling, postharvest treatments, **cold chain management** including control of storage temperature, relative humidity and gaseous composition are essential to minimize physiologica activities, microbial spoilage and reduction in risk of foodborne illness. Alo with the improvement in conventional postharvest technologies, prioritizat of cold supply chains, nanotechnology, smart packaging, irradiation, impro storage techniques, and non-thermal novel processing technologies incl cold plasma technique, pulsed electric fields (PEF), ultrasonication, high pr processing (HPP), vacuum drying and vacuum frying can be helpful in m decontamination without affecting nutritional and sensory characte foods, and in prolonging storage life of fruits and vegetables to re loss of harvested produce, which in turn will boost farmers' incom economy, and increase the affordability of fruits and vegetables for global malnutrition. Initiatives from both public and private sectors more on infrastructure development especially cold chains, the element in prolonging shelf life of perishables, and processing u more funds for postharvest research areas, and emphasising m building in postharvest management, and food standards for food safety standards can reduce postharvest losses remarka

Modified Atmospheric Packaging (MAP)

With the increase in demand for fresh and natural products without the addition of harmful chemicals, modified atmospheric packaging (MAP) is gaining popularity for extending the shelf life of fresh fruits and vegetables without much affecting their fresh-like appearance. Modified atmosphere packaging works on the concept of reducing the level of oxygen (O_2) inside the packing bag and at the same time increases the concentration of carbon dioxide (Veljovic *et al.*, 2017). Edible coatings composed of polysaccharides, proteins, lipids and composites are also applied directly on the food surface by dipping, spraying or brushing to create a modified atmosphere as alternative to the protective waxy coatings to provide a barrier to moisture, oxygen, and solute movement for prolonging storage life of whole as well as minimally processed fruits and vegetables.

The atmosphere inside a package can be modified either by passive or active means. In case of **passive modification**, the rate of change and the final gas composition in the package depend largely on both the packaged product and the permeability of the packaging material. Packaged fruits and vegetables consume the oxygen of the surrounding environment and release carbon dioxide during respiration resulting in reduced oxygen concentration over time, which may cause anaerobic respiration and fermentation, deteriorating the quality of packaged produce. Considering the limitations of the passive atmosphere modification in achieving the desired atmosphere very slowly and chances of uncontrolled levels of oxygen, carbon dioxide or ethylene with a detrimental effect on the quality of the packaged produce, active modification of the atmosphere by 'smart packaging' is gaining importance to provide a solution to this problem.

Smart Packaging

It is an emerging technology in the food packaging industry, which incorporates both active and intelligent packaging. Smart packaging can meet the demands for fresh and safe food in better way than traditional packaging by transforming packaging from a passive to active state. Smart packaging is a collective term used to group some emerging food packaging technologies referred to as 'active packaging', and 'intelligent packaging'. The purpose of dding active packaging functionality to a packaging system is to improve the rotective functional value *e.g.*, antimicrobial materials or oxygen absorbers. The urpose of adding intelligent packaging functionality to a packaging system is to nprove its epistemic value providing additional information about the product ., time temperature indicators or ripeness indication (Robertson, 2012). One nificant area of technological overlap with active and intelligent packaging is t of nanotechnology applied to the food packaging industry. Nanotechnology platform technology and can be utilized in packaging to achieve the aims of rt packaging but has potential to pose health risks (Alfadul and Elneshwy,

2010). It must also be noted that the use of nanotechnology in food processing or packaging has the potential to cause significant consumer concerns (Siegrist *et al.*, 2007). The main concepts and types of active and intelligent food packaging, focusing on recent progress and new trends using biodegradable and biobased polymers were also described by Salgado *et al.* (2021). For commercial success, food products utilizing innovative active or intelligent packaging technologies must be accepted by consumers. Different innovative packing systems are shown in Figure 2.1.

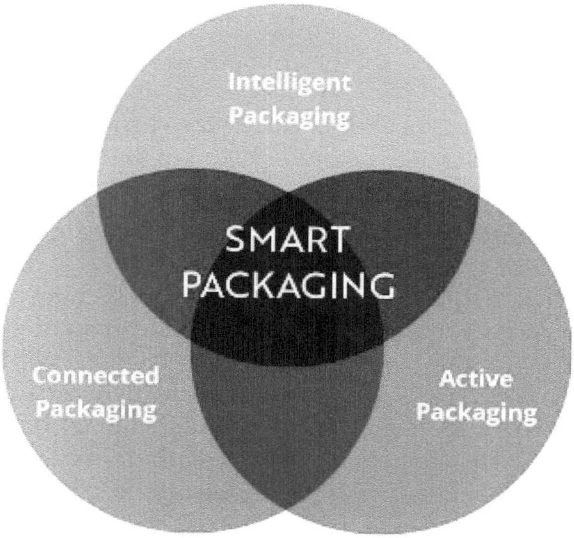

Figure 2.1: Innovative Packaging Systems.

Active Packaging

In active packaging, atmospheric modification is usually accomplished by creating a vacuum and then incorporating the desired gas mixture in the package. Compared to the passive method, active atmospheric modification is an instant method and takes place at the beginning of storage. The atmosphere then remain practically unchanged, provided that the proper barrier material is used and there is no leakage through pinholes or poor seals. Three main gases used in MAP are nitrogen (N_2), oxygen (O_2), and carbon dioxide (CO_2). Nitrogen is primarily used to displace oxygen and prevent package collapse. A low oxygen concentration is maintained for the packaged fruits and vegetables in order to sustain their basic process of aerobic respiration. And carbon dioxide is used for its bacteriostatic effect and the ability to slow down the respiration of many products. However, the optimum level of each gas for each commodity must be determined and used in order to maximize the positive effects. In some cases, additional gases are used in combination with the above mentioned gases.

Carbon monoxide (CO) is sometimes added to inhibit microbial growth but due to toxicity, its application is limited. Similarly, sulfur dioxide (SO_2) may be used to prevent oxidative browning and to control the growth of bacteria and moulds. Ethanol has also been used to enhance the firmness of tomatoes, improve flavour, and reduce fungal activity. Active packaging techniques for extending shelf life, maintaining quality and safety of foods can be divided into two categories; absorbing system (scavengers) and releasing system (releasers). Absorbing (scavenging) systems remove undesired compounds such as oxygen, carbon dioxide, ethylene, excessive water, taints *etc.* Releasing systems actively add or emit compounds like carbon dioxide, antioxidants and preservatives to the packaged food or into the head-space of the package. Depending on the physical form of active packaging systems, absorbers and releasers can be a sachet, label or film type. Sachets are placed freely in the head-space of the package and labels are attached into the lid of the package avoiding direct contact with food.

The most common ethylene (C_2H_4) scavenging agent is potassium permanganate ($KMnO_{4)}$ which in presence of water, oxidizes C_2H_4 to CO_2 and H_2O (Gaikwad *et al.*, 2020). Warsiki (2018) developed chitosan films activated with $KMnO_4$ to wrap tomatoes that managed to prolong their shelf-life. Nevertheless, $KMnO_4$ cannot be utilized in direct contact with foods due to its high toxicity and has a limited long-term efficacy in high-moisture environments (Yildirim *et al.*, 2018). Ethylene can also be eliminated by physical adsorption on active surfaces such as activated carbon, carbon impregnated with metal catalysts or activated clays like zeolite, vermiculite and montmorillonite incorporated into the packaging film. The use of 1-methylcyclopropane (1-MCP) is another option to diminish the effect of C_2H_4. Its action mechanism implies its irreversible union to the C_2H_4 receptor in fruit, thus blocking C_2H_4 effects by competitive inhibition. It is encapsulated in cyclodextrins in commercial products and releases when the formulation is mixed with water. Ortiz *et al.* (2013) showed that pillows prepared with soybean proteins could use the relative humidity inside the package to release 1-MCP and delay tomatoes softening and suggested that it could be useful for postharvest treatments in the own package.

Intelligent Packaging

Intelligent packaging is one type of smart packaging that may include sensors for detecting the freshness or temperature of the packaged contents. Freshness or time/temperature indicators or sensors can be integrated inside or outside the packaging to show retailers and customers if an important limit has been exceeded. An intelligent package can carry out many intelligent functions, like detecting, registering, locating, communicating and applying scientific logics to help in decision-making, extending shelf-life, maintaining quality, improving safety, providing information and indicating possible problems with the help of sensors. Unlike active packaging, intelligent packaging is not aimed at releasing components onto the food or headspace to act directly for extending shelf-life of food, instead, it conveys information about food quality to manufacturers, handlers, retailers, and consumers.

The components, which are attached as labels or incorporated or printed on the food packaging material, offer better possibilities to check product quality, track critical items and provide more detailed information during all the food supply chain including storage, transport, distribution and sale (Lee *et al.*, 2015). They can also inform about product history such as storage conditions, composition of headspace, microbial growth, *etc.* (Realini and Marcos, 2014). Halonen *et al.* (2020) described the development of biobased antimicrobial films and sensors to be used as indirect indicators of food spoilage.

Time-Temperature Indicators (TTIs)

TTIs are typically small self-adhesives for attaching, or small devices for incorporating into the package that provide a visual indication, most commonly by changing colour or moving colour front, as a response to temperature history during distribution and storage. These indicators may be placed in the shipping container or in the individual consumer package. There are three basic types of TTIs- full history (those that start integration immediately once activated), partial history (those that start integration when a certain temperature threshold is achieved), and critical temperature or abuse indicators (that show a change if a certain temperature is achieved and whose response does not depend on the time the product has been packaged). The major mechanisms on which the TTIs are based include enzymatic reaction, polymerisation, melting point and diffusion of a substance. In the first case, once the indicator is activated, an enzymatic reaction takes place causing a pH change and a consequent colour change. Indicators that work based on polymerisation have a polymer coating that gets darker at a rate that depends on the temperature. The colour of the polymer is compared to a colour reference surface printed on the indicator. These indicators are usually self-activated, thus requiring deep low temperature storage before use. Three commercially available TTIs namely MonitorMark, CheckPoint and Fresh-Check work on the principle of diffusion, enzymatic reaction and solid state polymerisation respectively. The TTIs in the fruits and vegetables sector have an interesting application in minimally processed products, such as fresh-cut fruits, vegetables and salads, since these fresh-cut products require a very good temperature control for maintaining quality and safety.

Connected Packaging

It is an innovative packaging, finds applications in almost all retail product categories, including toys, cosmetics, games and clothing. It allows consumers to interact with a product through a code on the package that can be activated with a mobile device. Connected packaging generally used by brands to deliver exclusive content to its customers, can also be used for high value fruits and vegetables to transform simple packaging into digital experience, through a mobile phone tap on a QR code.

Nanotechnology in Packaging

Application of **nanotechnology** in postharvest management of horticultural crops by nanocomposite packaging, nanoencapsulation and nanoemulsion coating of fruits and vegetables can be useful in extending shelf-life of perishables by delaying ripening and/or inhibiting the growth of pathogenic microorganisms. Nanocomposites are materials that incorporate nanoparticles into a matrix of base material. Polyvinyl alcohol (PVA), polyethylene terephthalate (PET), polystyrene (PS)) and polyethylene (PE) are some of the polymer matrices generally used in preparation of nanocomposites. The polymer materials are reinforced with nano sized particles of silver (Ag), titanium dioxide (TiO2), silicon dioxide (SiO2) also known as silica, and clay or layered silicate to develop nanocomposites. The size of the reinforced material has direct impact on the property of the nanocomposites, as extremely high surface to volume ratio of nanoparticles dramatically changes their properties. Among various applications of nanocomposites, one use as oxygen and gas barrier is very important, and in food packaging, these composite materials are used for blocking CO2 and water vapour, and removal of ethylene from the packaging material to enhance the shelf life of fruits and vegetables. Electrostatic interactions between nanocomposite and bacterial cell wall or fungal cell membrane destroy the cell structure and increase the cell membrane permeability leading to the leakage of intracellular substances and impairment of vital microbial activities. Nanoparticle based packaging has many advantages over conventional packaging in terms of antimicrobial properties, barrier properties and greater strength. Nano-titanium dioxide (TiO_2) with photocatalytic capabilities can oxidize ethylene into carbon dioxide and water, thus preventing loss due to excessive ripening. Sami *et al.* (2021) reported that the chitosan/nano-silica with the addition of nisin coating extended shelf-life of fresh-cut cantaloupe up to 8 days at 4°C by forming semi-films against microorganisms and enzymes. Nano-silicon oxide (SiO_2) coating has also been used in apples for extending shelf life.

Transportation

An ongoing effort has to be made by both public and private sectors to implement innovative measures for adding value to the transportation sector and making it capable of appropriately responding to the growers' demands. Due importance should be given to transport systems for transportation of fruits and vegetables to local market, packing house, processing units, store house, long distant market, and also for export by air and sea transport. Refrigerated containers and trailers need to be customarily used for long distance transportation by sea, rail or truck in extending shelf life and preserving the quality of fresh fruits and vegetables. In spite of high initial investment and operating cost, only cold supply chains can provide basic requirements including proper control of temperature, humidity and ventilation during transportation. In addition, fresh produce should be properly packed and stacked using pallets to avoid compression damage due to excessive movement or vibration. Cold chain model needs to be designed using refrigeration components including pre cooling chiller, cold room, and truck refrigeration. Considering the importance of cold supply chains in prolonging shelf life and reducing postharvest losses of horticultural crops, investment in cold chain is increasing in India. It is estimated that India's investment in cold chain will be over Rs. 11,100 Crores (about USD 15 billion) over the next five years.

Storage

In developing countries, especially in tropical countries, long term storage of fruits and vegetables is a major problem where sufficient cold stores are not available. Along with the microbial spoilage of fruits and vegetables, sprouting in tubers and bulbous crops is a major problem during storage in conventional low cost storage structures. Therefore, proper storage facilities with the provisions for modification of temperature, relative humidity, gaseous concentration and pressure as per the nature of commodity need to be created for prolonging storage of fruits and vegetables. Several improved storage structures have been developed with the manipulation of temperature, humidity, gaseous concentration and atmospheric pressure for shelf life extension of fruits and vegetables.

Cold Storage

Cold storage or low-temperature storage, the enclosed chamber with electric refrigeration unit is the most commonly adopted improved technique for long-term storage of fruits and vegetables. Most of the cold storages are equipped with full control of temperature, and can be adjusted ranging from 0–10 °C or more for a number of commodities.

Controlled Atmosphere (CA) Storage

In CA storage, the concentration of oxygen and carbon dioxide is manipulated and constantly regulated along with the maintenance of low temperature and humidity during storage. This is an advance method of storage in which atmospheric composition is different than the normal atmosphere where oxygen level is generally kept below 8 per cent and CO_2 level is kept above 1 per cent combined with refrigeration. Despite the fact that the reduced concentration of oxygen slows down metabolism, it also affects the ethylene production, since oxygen is necessary for transforming 1-amino cyclopropane -1-carboxylic acid to ethylene. In the horticultural industry, CA storage is most successful invention to prolong the shelf life of fresh produce (Bodbodak and Moshfeghifar, 2016), which was first reported by Kidd and West (1927) with its commercial application in pome fruits. Several studies showed the positive effects of low oxygen and elevated level of carbon dioxide on shelf life and quality of fruits stored in CA storage (Table 2.1).

Table 2.1: Effect of Various CA Storage Conditions on Quality of Fruits

Crop	*Concentrations*		*Observations*	*References*
	O_2	*CO_2*		
Apple (Granny Smith)	<0.5 per cent	1 per cent	Retained firmness for a long time	Mditshwa *et al.* (2018)
Strawberry	5 kPa	15 kPa	Increased strawberry shelf-life by 3 days (based on disease incidence)	Alamar *et al.* (2017)
Acovado	2.5 and 5 kPa	5 and 7.5 kPa	Optimized shelf life and fruit quality	Basuki *et al.* (2016)
Litchi	1 per cent	5 per cent	Reduced weight loss	Ali *et al.* (2016)
Mango (cv. 'Kensington Pride' and 'R2E2'	3 per cent	6 per cent	Extended the storage life with good quality up to 6 weeks	Singh (2015)
	At 13°C			
Pomegranate	10 kPa	5 kPa	Reduced weight loss	Pareek *et al.* (2015)
	5 kPa	5 kPa	Minimized fungal decay	
Plum (cv. Laetitia)	1-2 kPa	1-3 kPa	Delayed ripening	Steffens (2014)
Mango (cv. Delta R2E2)	3 per cent	6 per cent	Reduced respiration rate, and delayed de-greening of the skin and fruit softening. Increased TSS, total sugars content and taste of the ripe fruit	Lalel *et al.* (2005)

Hypobaric Storage

Hypobaric storage, also known as low pressure storage is a system of storing harvested commodities while ventilating with air at less than atmospheric pressure. This condition helps in increasing shelf life, decay incidence and maintain firmness of the commodity by reducing metabolic activities. Storage of harvested produce by application of reduced pressure was first used by Workman *et al.*, in 1957, when they stored tomato fruits and recorded reduced respiration rate at 20°C temperature under 88 mm Hg storage as compared to those stored at 20°C under atmospheric pressure. It is very essential for storing crop in hypobaric storage that storage atmosphere is constantly being changed in order to maintain required oxygen level, since the produce in hypobaric storage is continuously respiring. This is achieved by vacuum pump evacuating the air from the store and the store atmosphere is continuously being replaced from the outside. The inlet and outlet of air from the store should be balanced to achieve the desired pressure in the store for a particular commodity. The reduced pressure inside the store can result in rapid water loss from the crop, to overcome this problem, the relative humidity inside it must be cent per cent or near to the saturation point of air (Thompson, 2016). Hypobaric storage is effective in reducing decay incidence, controlling browning and imparting better postharvest life in the fruits and vegetables. Effect of hypobaric storage on the quality and shelf life of some fruits is shown in Table 2.2.

Table 2.2: Effect of Hypobaric Storage on the Quality and Shelf Life of some Fruits

Crop	Storage Pressure	Observations	References
Strawberries	380 mm Hg, at 20°C or 5°C,	Increased the activity of defence related enzymes, and delayed rot development	Hashmi *et al.* (2013)
Jujube cv. Lizao	20.3, 50.7 or 101.3 kPa	Delayed the loss of ascorbic acid, and prolonged the postharvest life	Cui (2008)
Loquat	40-50 mm Hg	Reduced decay and fruit browning	Gao *et al.* (2006)
Avocado	61–81 mm Hg	Increased fruit firmness up to 30 days	Burg(2004)

Irradiation

Food irradiation is a physical means of food preservation that involves exposure of pre-packaged or bulk foodstuffs to ionizing radiations including gamma rays, X-rays, or electrons, which kills most of the spoilage causing organisms, inactivates the enzymes responsible for browning, and also delay the ripening of fruits and sprouting of vegetables. With an objective to reduce the use of chemicals in preserving fruits and vegetables, the non-residual feature of ionizing radiation has gained popularity in many countries. Ionizing radiation from cobalt-60 or caesium-137, or machine-generated electron beams are used as a source of irradiation for extending the shelf life of fresh produce. The energy from the radiation breaks the bonds in the DNA of microbes, causing defects in the genetic makeup and making them unable to reproduce. Ferrier (2010) reported that irradiation is used for the control of potato and onion sprouting, and strawberry decay commercially. While much of the focus of irradiation use on fresh fruits and vegetables for extending shelf life and reducing decay, it has been considered as an effective means for killing, sterilizing or preventing further spread of foreign invasive pests of quarantine importance across international borders. Gamma irradiation has also been employed for decontamination or sterilization of dehydrated fruits and vegetables. However, the irradiation being an expensive method, is generally used for the storage of high-value fresh commodities or for export.

Emerging Processing Technologies

Although conventional processing techniques such as drying, canning, bottling, pickling and freezing are often used for extending shelf life, and to

Table 3: Applications of Food Irradiation

	Application	*Dose (kGy)*
Low dose (up to 1 kGy)	Inhibit sprouting (potatoes, onions, yams, garlic)	0.06 - 0.2
	Delay in ripening (strawberries, potatoes)	0.5 - 1.0
	Prevent insect infestation (grains, cereals, coffee beans, spices, dried nuts, dried fruits, dried fish, mangoes, papayas)	0.15 - 1.0
	Parasite control and inactivation (tape worm, trichina)	0.3 - 1.0
Medium dose (1 kGy to 10 kGy)	Extend shelf-life of raw and fresh fish, seafood, fresh produce	1.0 - 5.5
	Extend shelf-life of refrigerated and frozen meat products	4.5 - 7.0
	Reduce risk of pathogenic and spoilage microbes (meat, seafood, spices, and poultry)	1.0 - 7.0
	Increased juice yield, reduction in cooking time of dried vegetables	3.0 - 7.0
High dose (above 10 kGy)	Hygienization of spices	10.0
	Sterilization of spices, dry vegetable seasonings	30.0 max
	Sterilization of packaging material	10.0 - 25.0
	Sterilization of foods (hospitals)	44.0

Source: Fellows, 2018; Xuetong, 2018.

add value to fruits and vegetables, nowadays, considering the loss of nutritional quality in thermal processing, a number of non-thermal novel processing techniques including cold plasma, pulsed electric field (PEF), high-pressure processing (HPP) and ultrasonication are available to extend the shelf life of fruits and vegetables.

Cold Plasma (CP) Technology

Cold plasma technology is one of the emerging non-thermal processing techniques attracting more interest from researchers and food processors to prolong the shelf life of fruits and vegetables. Cold plasma technique can be used for inactivation of microorganisms, removal of toxins, degradation of pesticides and dyes, and packaging modification. Plasma, the fourth state of matter exists as an ionized gas containing electrons, free radicals, ions, *etc.* possessing a net neutral charge. It is induced under varying temperatures and pressures by energizing a neutral gas, and classified as thermal and non-thermal plasma. As the thermal technology has disadvantage of altering food quality, postharvest treatment with cold plasma, a gas-derived mix of atoms, excited molecules and charged particles can extend the shelf life without any adverse effect on fresh produce. It is reported that atmospheric plasma at 18 kV can kill *Salmonella spp.* on strawberry samples by 1.7 - 2.3 log CFU/sample. Likewise, using dielectric barrier discharge (DBD) plasma at 80 kV for 5 min successfully decontaminated cherry tomatoes containing *Escherichia coli, Salmonella typhimurium,* and *Listeria monocytogenes* by up to 3.5, 3.8, and 4.2 log CFU/tomato respectively. Therefore, sterilization using cold plasma at low cost with no or minimal impacts on the quality attributes like colour, texture, sensory and nutritional value of treated products can be an effective alternative to heat-based processing techniques.

Pulsed Electric Fields (PEF)

Though the thermal processing is generally used for microbial decontamination of food, it is not a preferred method due to loss of nutritional quality and sensory properties of food. Therefore, in addition to nanotechnology, irradiation and cold plasma technology, pulsed electric fields (PEF), another novel non-thermal decontamination technology is now being considered as an effective tool for food decontamination, reducing food borne illness and shelf life extension of food by applying short electric pulses with very little or no detrimental effect on quality attributes of food. Application of pulsed electric fields of high intensity, created from high voltage pulses generated by high voltage pulse generator for a very short duration from a few microseconds to milliseconds causes permeabilization or puncturing of cell membranes triggering death of microorganisms. PEF food processing system is illustrated in Figure 2.2.

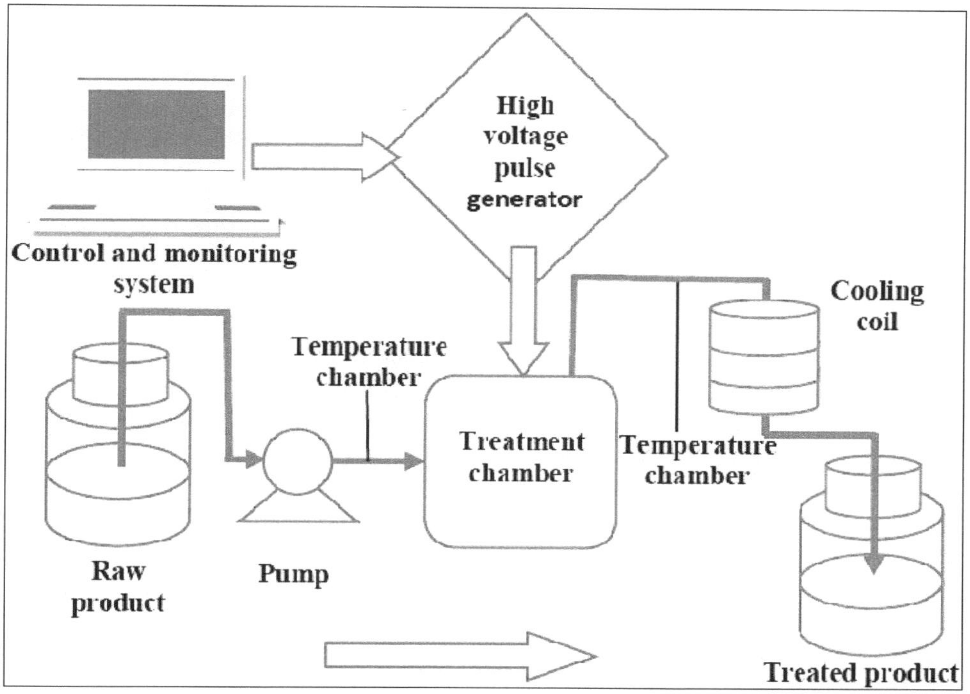

Figure 2.2: Flow Chart of a PEF Food Processing System with Basic Components (*Source*: Maged and Ayman, 2012).

Food products with no air bubbles and low electrical conductivity are best suited for processing by pulsed electric fields technology. In addition to various liquid foods like milk, yogurt, soups *etc.*, different fruit juices such as apple juice, orange juice, tomato juice, carrot juice, and some other foods like apple sauce, pea soups and salad dressings have been found to maintain sensory attributes and nutritional quality for an extended period after processing by this novel method using pulsed electric fields without application of heat or any preservative. The effects of pulsed electric fields (PEF) on biomembranes have been thoroughly studied by scientists since the use of PEF has attracted great interest in several scientific areas such as cell biology, biotechnology, medicine and food technology.

High Pressure Processing

The growing concern about food borne diseases and intense use of preservatives and chemicals has made the consumers to shift towards natural and healthy foods such as fruits and vegetables and their products in terms of natural flavour and taste, and free from additives and preservatives. This demand has triggered the need for the development of a number of non-thermal processing techniques, of which high-pressure processing technology has proven to be very valuable. High pressure processing commonly known as **HPP** technology, and sometimes termed as hydrostatic pressure processing (HPP) is a cold pasteurization or non-thermal processing technique in which food, sealed in flexible and water-resistant package is subjected to a high pressure of different ranges transmitted by water known as hydrostatic pressure usually from 400 MPa to 600 MPa (megapascal) in metric system or 58000 psi to 87,000 psi (pounds per square inch) in the imperial system for a few seconds to a few minutes, usually one to five minutes depending on the nature of food and the target. The food product to be treated at high pressure is packaged in a flexible container and placed inside a pressure vessel submerged in water or a liquid medium, which transmits the pressure. These high pressures to packaged fruit and vegetable products in high pressure processing equipment kill most of the harmful microorganisms, such as salmonella, E. coli and Listeria monocytogenes by damaging their cell components without altering colour, flavour, texture or nutritional value of the product. As in this pressure processing method, no chemical preservative is used, it is also referred as natural pressure cold pasteurization. HPP is an effective alternative to heat processing for fruit and vegetable products such as juice, jam, puree, sauce *etc.* in extending shelf life and preserving nutritional and sensory quality. Several commercial products, including fruit juices of mandarin orange, grapefruit, apple and sweet orange, and carrot juice and broccoli-apple juice mixture treated with HHP are currently available in different markets. But, the main limitation in using this technique with high pressure treatment is to safely and repeatedly generate high pressure in a vessel, solid foods have to be processed in batch or semi continuous equipment. In spite of relatively high capital cost of equipment, the use of high pressure treatments and the range of products on which the treatments are applied continue to grow every year (Gopal, *et al.*, 2017).

Ultrasonication

Ultrasonication is one of the emerging novel technologies, now used in the field of food science and technology. Ultrasound being non-toxic and environment friendly is considered as green technology, which saves lot of energy and maximizes intended output. Among diverse applications, ultrasound has been employed in studying composition of fruits, vegetables, and dairy products, and detecting microbial contamination in dairy and canned foods. Ultrasonication has also gained huge application in the food industry such as processing, extraction, emulsification, homogenization and preservation. It is used to maintain postharvest quality attributes in fresh fruits and vegetables including sensory characteristics, nutritional value and microbial safety, and can be a substitute for washing of fruit and vegetable in food industry. Use of ultrasonication in guava juice, orange juice and tomato juice has been reported to maintain fresh quality, nutritional value, and microbiological safety. Ultrasonication cleaners (20–400 kHz) have been efficiently used to produce fruits and vegetables free of contamination (Lin and Erel, 1992) and at 40 kHz, it has been applied on strawberry fruits in which decay and infection was considerably reduced along with maintenance of quality (Cao *et al.*, 2010). Principle of microbial inactivation by ultrasonic waves is demonstrated in Figure 2.3.

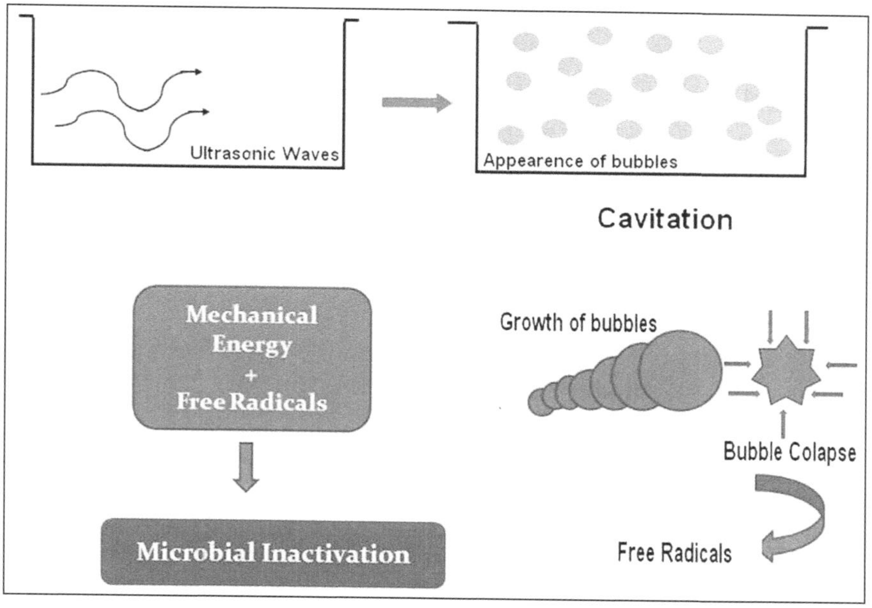

Figure 2.3: Cavitation Phenomenon and Microbial Inactivation by Ultrasonic Waves (*Source*: de Sao Jose *et al.*, 2014).

Along with securing food for all, food safety also needs to be focused, as it is directly related with human health. Although many techniques have been developed, no method is equally suitable for all the commodities due to wide variations in nature and texture of fruits and vegetables, so recommendations should be crop, variety and location specific. Efforts should also be made for identifying and developing varieties having long shelf life and good processing quality. Coordinated activities by growers, handlers, storage operators, processors, wholesalers and retailers are required to maintain quality and reduce postharvest loss of perishable commodities.

References

Alamar, M. C., Collings, E., Cools, K. and Terry, L. A. 2017. Impact of controlled atmosphere scheduling on strawberry and imported avocado fruit. *Postharvest Biol. Technol.*, **134:** 76-86.

Alfadul, S. M. and Elneshwy, A. A. 2010. Use of nanotechnology in food processing, packaging and safety- review. *African J. Food, Agric. Nutr. Dev.*, **10** (6): 1-21.

Ali, S., Khan, A. S., Malik, A. U. and Shahid, M. 2016. Effect of controlled atmosphere storage on pericarp browning, bioactive compounds and antioxidant enzymes of litchi fruits, *Food Chem.*, **206**: 18–29.

Basuki, E., McGlasson, W. B. and Skurry, G. 2016. Polyamines association in textural changes of avocado during controlled atmosphere storage. *Imp. J. Interdiscip. Res.*, **10** (2): 1716-1721.

Bodbodak, S. and Moshfeghifar, M. 2016. Advances in controlled atmosphere storage of fruits and vegetables. **In:** *Eco-Friendly Technology for Postharvest Produce Quality*, Academic Press, Elsevier, pp. 39-76.

Burg, S. P. 2004. *Postharvest Physiology and Hypobaric Storage of Fresh Produce.* CABI, Wallingford, UK.

Cao, S., Hu, Z., Pang, B., Wang, H., Xie, H., and Wu, F. (2010). Effect of ultrasound treatment on fruit decay and quality maintenance in strawberry after harvest. *Food Control,* **21**: 529–532.

Cui, Y. 2008. Effect of hypobaric condition on physiological and biochemical changes of Lizao jujube. *J. Anhui Agric. Sci.,* **36** (29): 12900-12901.

de Sao Jose, J. F. B., de Andrade, N. J., Ramos, A. M., Vanetti, M. C. D., Stringheta, P. C. and Chaves, J. B. P. 2014. Decontamination by ultrasound application in fresh fruits and vegetables. *Food Control,* **45:** 36–50.

Ferrier, P. 2010. Irradiation as a quarantine treatment. *Food Policy.* **35**: 548–555.

Gaikwad, K. K., Singh, S., and Negi, Y. S. 2020. Ethylene scavengers for active packaging of fresh food produce. *Environ. Chem. Lett.* **18**: 1–16.

Gao, H. Y., H. J. Chen, W. X. Chen, Y. T. Yang, L. L. Song, Jiang, Y. M. and Zheng, Y.H. (2006). Effect of hypobaric storage on physiological and quality attributes of loquat fruit at low temperature. *Acta Hortic*, **712**: 269–274.

Gopal, K.R., Kalla, A.M. and Srikanth, K. 2017. High Pressure Processing of Fruits and Vegetable Products: A Review. *Int. J. Pure App. Biosci.* **5** (5): 680-692.

Halonen, N. J., Palvölgyi, P. S., Bassani, A., Fiorentini, C., Nair, R., Spigno, G., *et al.*, 2020. Bio-based smart materials for food packaging and sensors–a review. *Front. Mater*, **7**: 82. doi: 10.3389/fmats.2020.00082.

Hashmi, M. S., East, A. R., Palmer, J. S. and Heyes, J. A. 2013. Pre-storage hypobaric treatments delay fungal decay of strawberries. *Postharvest Biol. Technol.*, **77**: 75–79.

Lalel, H. J. D., Singh, Z. and Tan, S. C. 2005. Controlled atmosphere storage affects fruit ripening and quality of 'Delta R2E2'mango. *J. Hortic. Sci.*, **80** (5): 551-556.

Kidd, F. and West, C. 1927. A relation between the concentration of oxygen and carbon dioxide in the atmosphere, rate of respiration, and length of storage of apples. *Food Investigation Bd. Rpt.* London, pp.41-42.

Lee, S. Y., Lee, S. J., Choi, D. S., and Hur, S. J. 2015. Current topics in active and intelligent food packaging for preservation of fresh foods. *J. Sci. Food Agric.* **95**: 2799–2810.

Lin, I., and Erel, D. 1992. Dynamic ultrasonic cleaning and disinfecting device and method. US Patent No. 5113881A. Washington, DC: U.S. Patent and Trademark Office.

Maged, E. A. M. and Ayman, H. A. E. 2012. Pulsed Electric Fields for Food Processing Technology. **In**: *Structure and Function of Food Engineering*. A. A. Eissa (Ed,), Intech Open.

Mditshwa, A., Fawole, O. A. and Opara, U. L. 2018. Recent developments on dynamic controlled atmosphere storage of apples- A review. *Food Packag. Shelf Life*, **16**: 59-68.

Ortiz, C. M., Mauri, A. N., and Vicente, A. R. 2013. Use of soy protein based 1-methylcyclopropene-releasing pads to extend the shelf life of tomato (*Solanum lycopersicum* L.) fruit. *Innov. Food Sci. Emerg. Technol*, **20**: 281–287.

Pareek, S., Valero, D. and Serrano, M. 2015. Postharvest biology and technology of pomegranate. *J. Sci. Food Agric.*, **95** (12): 2360–2379.

Realini, C. E., and Marcos, B. 2014. Active and intelligent packaging systems for a modern society. *Meat Sci.* **98**: 404–419.

Robertson, G. L. 2012. *Food Packaging: Principles and Practice*. CRC Press, Massachusetts, MA.

Salgado, P. R., Di Giorgio L., Musso, Y. S. and Mauri, A. N. 2021. Recent developments in smart food packaging- focused on biobased and biodegradable polymers. *Front. Sustain. Food Syst.* **5**: 630393. doi: 10.3389/fsufs.2021.630393.

Sami, R., Almatrafi, M., Elhakem, A., Alharbi, M., Benajiba, N. and Helal, M. 2021. Effect of nano silicon dioxide coating films on the quality characteristics of fresh-cut cantaloupe. *Membranes,* **11** (2): 140, doi: 10.3390/membranes11020140.

Siegrist, M., Cousin, M. E., Kastenholz, H. and Wiek, A. 2007. Public acceptance of nanotechnology foods and food packaging: The influence of affect and trust. *Appetite,* **49** (2): 459-466.

Singh, Z. 2015. Controlled atmosphere storage of mango fruit: Challenges and thrusts and its implications in international mango trade. *Acta Hortic.,* **1066**: 179-191

Steffens, C. A., Amarante, C. V., Alves, E. O. and Brackmann, A. 2014. Fruit quality preservation of 'Laetitia' plums under controlled atmosphere storage. *An. Acad. Bras. Cienc,* **86** (1): 485-494.

Thompson, A. K. 2016. *Fruit and vegetables-harvesting, handling and storage,* 3rd ed. Oxford, UK: Wiley Blackwell.

Veljoviæ, S., Boonsiri, K., Maina, M., Semavor, E., and Mitra, S. 2017. Infection capacity of the pathogens *Penicillium italicum* and *P. expansum* in orange during storage. *Food and Feed,* **44** (1): 31-38.

Workman, M., Pratt, H. K. and Morris, L. L. 1957. Studies on the physiology of tomato fruit. I. Respiration and ripening behaviour at 20 °C as related to date of harvest. *Proc. Am. Soc. Hortic. Sci,* **69**: 352–365.

Warsiki, E. 2018. Application of chitosan as biomaterial for active packaging of ethylene absorber. *IOP Conf. Ser. Earth Environ. Sci.,* **141**: 12036.

Yildirim, S., Röcker, B., Pettersen, M. K., Nilsen-Nygaard, J., Ayhan, Z., Rutkaite, R. 2018. Active packaging applications for food. *Compr. Rev. Food Sci. Food Saf.,* **17**: 165–199.

Chapter 3

Issues in Quality and Safety of Fresh and Fresh-cut Produce

Introduction

Fresh fruits and vegetables are major sources of macronutrients such as dietary fibre, a part of carbohydrate, which is absent in animal foods, and micronutrients such as minerals and vitamins like ascorbic acid, thiamine, riboflavin, vit. B-6, niacin, folate, pro-vit. A, vit. E *etc.* Some Fruits and vegetables are rich in protein, and most of the fruits and vegetables contain low levels of fats. These horticultural crops are universally considered as the important part of the human diet. The edible portion of the most types of fruits and vegetables usually contains 70-90 per cent of water, and generally are low in calories with a few exceptions. Fruits and vegetables provide all the six major classes of nutrients of food like carbohydrates, proteins, fats, minerals, vitamins, and water. Regular consumption of plenty of fruits and vegetables can reduce the chances of heart disease and stroke, control blood pressure and cholesterol, prevent some types of cancer, and guard against cataract and macular degeneration, two common causes of vision loss. They also play key role in neutralizing the acids produced during digestion of proteinaceous and fatty foods, and also provide valuable fibres to digest food properly and prevent constipation. Some of the fruits and vegetables are very good sources of carbohydrates *e.g.* apple, apricot, banana, mango, jackfruit, oranges, pineapple, grapes, potato, sweet potato, cassava, onion, garlic *etc.* Fruits and vegetables rich in protein are almond, walnut, cashew nut, guava, lemon, apricot, peas, soybean, winged bean, lima bean, asparagus, broccoli, spinach and other leafy vegetables Orange-fleshed and purple-fleshed sweet potatoes are rich in β-carotene and anthocyanins respectively. Increased consumption of orange-fleshed sweet potatoes having high β-carotene can contribute considerably to alleviate dietary deficiency of vitamin A, and pigmented cultivars of purple-fleshed sweet potatoes having high anthocyanins and total phenols, and orange-fleshed sweet potatoes rich in β-carotene can also act as good sources of antioxidant (Mitra *et al.*, 2021). Most of the fruits and vegetables contain low levels of fats. Although a few fruits and vegetables are high in fat, such as avocados, olives, cashew nut, coconut and soybean, most of them contain high levels of monounsaturated and polyunsaturated fats hence their nutritional benefit remains high.

Some of the fruits and vegetables rich in protein, vitamins, minerals and antioxidants, and low in carbohydrates, fats and anti-nutrients seem to help in reducing detrimental effects of high fat and high sugar diets to the hippocampus, a major component of the brain. Being protective food, different fruits and vegetables are gaining popularity in almost all over the world as nutrition-security crop to fight oxidative stress, improve general cognition, reduce cholesterol levels and help in maintaining healthy blood pressure levels (Mitra, 2021).

Enhanced awareness of good nutritional value is a leading factor contributing to the increased consumption of fresh fruits and vegetables. However, along with the interest in nutritional value, concern about the quality and safety of fresh fruits and vegetables as well as fresh-cut produce has been increasing because of recent findings on the high level of diet-related health problems in children and adults.

Quality

Quality refers to the combination of attributes of a commodity having significance in determining the degree of acceptability of the produce to a consumer, *i.e.* quality is a measure of the degree of excellence or degree of acceptability by the consumers. The quality of fresh fruits and vegetables may be explained in terms of fresh market quality, table quality, keeping quality, appearance quality, nutritional quality and processing quality. Different parameters such as sensory properties like appearance, texture, taste and aroma, nutritional values, chemical composition, mechanical properties, functional properties and free from defects are considered for evaluating quality of a produce.

The International Organization for Standardization (ISO) defines quality as "the totality of features or characteristics of a product that bear on its ability to satisfy the stated or implied needs." Webster defines quality more simply as the "degree of excellence." Some aspects of quality can be measured, such as purity, colour, maturity, ripeness, nutritional value, *etc.* Other characteristics are highly subjective, such as flavour, which can only be assessed by the person doing the tasting. Aroma, an important component of flavour and quality perception, can be measured for the presence of volatiles, but consumers vary in their sensitivity to these volatiles. In commerce, the "quality" of fresh produce as it applies to pricing is influenced strongly by supply and demand. Understanding that quality is made up of many factors that are often subject to interpretation that varies from person to person highlights the complexity of defining quality.

Quality Attributes, Grades and Standards

Quality includes some attributes that are subjective. For business purposes, uniformity of produce quality is important, and it is essential to have practical, objective grades and standards that define quality. It is not possible to clearly differentiate each quality attribute from all others as they all relate to each other in some manner. However, experts have classified some quality characteristics for the purpose of evaluating them in the course of conducting business and for scientific studies. These are grouped in three general categories namely external, internal and hidden attributes.

External Attributes

These are the ones that are perceived immediately when the product is encountered. These are the attributes that are seen and felt. Appearance quality includes colour, size, shape, and the presence or absence of defects. Defects may be due to several causes such as insect injury, environmental factors (sunburn, splitting from excessive watering, *etc.*), handling injury (bruising, cuts, scrapes, *etc.*) or decay. Aroma may be sensed externally but is more commonly measured as an internal characteristic of the flesh. Collectively, the external attributes are the factors most likely to influence the consumer's decision to purchase or reject a product.

Internal Attributes

These typically are not obvious until the product is cut or bitten, although a trained evaluator may make an accurate prediction of internal quality based on external characteristics. For example, a watermelon might appear to be ripe to an average customer in the supermarket, but for a trained observer there may be external characteristics that indicate the fruit will not have good eating quality. Aroma, flavour, texture, colour, turgidity and firmness are a few examples of internal quality characteristics. The combination of internal and external characteristics will determine if the customer is likely to make a second purchase of the product. For some fruits and vegetables the appearance quality may still be good long after flavour and other sensory quality attributes have deteriorated, resulting in disappointment for the consumer and reluctance to purchase the product in the future.

Hidden Attributes

Its include wholesomeness, nutritional value, and safety. As the name implies, these are almost impossible for the average consumer to assess. However, the perception of the hidden attributes may play a large role in the customer's decision to purchase. For example, wholesomeness and nutritional value are generally associated with an appearance of freshness. Items that are wilted or are not brightly coloured may not be perceived to be wholesome and may be rejected at the point of sale. Information about nutritional value sometimes is posted at the point of sale.

Quality of harvested fruits and vegetables deteriorates due to mechanical injury including bruising, cracking, cuts; microbial spoilage by fungi and bacteria; physiological reasons including respiration, transpiration, ethylene synthesis and ripening, and also due to enzymatic degradation.

As the maturity at harvest is one of the most important factors in determining final product quality and storage life, postharvest operations also include careful harvesting of a produce at proper stage of maturity with good quality that has positive impact on postharvest quality of fresh produce. The postharvest management is considered as a way to maintain quality, especially organoleptic quality, making the product more attractive to the consumer.

Due to great advancement in postharvest technology, maintenance of quality and safety of fresh as well as fresh-cut produce is now possible. Fresh fruits and vegetables continue their respiration and transpiration after harvesting, and the rates of respiration and transpiration are directly related to the shelf-life, thus these rates can be used to monitor deterioration during storage. The respiration and transpiration rates should be kept low as per the nature of the crop to maintain the fresh appearance, taste, aroma and other qualities of the harvested produce. The effect of temperature is tremendous on the rates of respiration, transpiration, decay control and both on production and sensitivity to ethylene. So the field heat should be reduced as soon as possible after harvesting by lowering the temperature and the fresh produce should be subjected to low temperature above the freezing temperature for non-chilling sensitive commodities and above the minimum safe temperature for chilling sensitive fruits and vegetables. Heat treatment is another view of temperature control. Postharvest heat treatments of fruits are currently used commercially in several countries for disease control, as a quarantine technology, to modify fruit responses to other stresses, but also for maintenance of fruit quality during storage. The impact on quality attributes varies with species, temperature and duration of the heat treatment (Paull and Chen, 2000; Pérez *et al.*, 2003). Manipulation of surrounding atmosphere composition of a perishable commodity is also an effective means in postharvest management. Modified atmosphere packaging and controlled atmosphere storage involve reduced oxygen and elevated carbon dioxide levels to prolong the postharvest life and to maintain quality of many fresh commodities through the reduction of the respiration rate. However, responses to atmospheric modification vary greatly among plant species, organ type and developmental stage, and duration and temperature exposure (Kader, 2002).

Changes in quality from harvest to consumption depend to a certain degree on the particular commodity, pre-harvest and postharvest factors and different treatments during handling and storage of fresh fruits and vegetables. A key component to successful marketing of food is the appearance of the product. Fresh fruits and vegetables are not sold if they do not look fresh and are plagued with soft and/or brown tissues. Preventing mechanical injury and browning will extend the shelf life of the fresh produce and increase its marketability.

Rejection of fruits and vegetables in international markets is also due to quality and safety issues. To avoid rejection from international markets during exports, fruits and vegetables should be free from non-authorized pesticides, pesticide residues exceeding permissible limits, inadequate labelling and packaging requirements, contaminants exceeding regulatory levels, and must be with the required nutritional information and/or with adequate general quality.

Quality Standards

Food standards consist of precise descriptors for the criteria that define the quality of the product. They provide common frames of reference that can be used as a basis for business transactions and for disputes to be settled by regulatory authorities. It is critical for the business community to have these grades and standards that provide uniformity in the determination of the quality of fresh produce. There must be a common language or terminology that both parties understand. This helps to establish market value and to prevent economic fraud. Without standards, the description of product quality could easily be misrepresented or misinterpreted. There are several types of standards in use today and all are based upon the various quality criteria. Official standards are those set by governments and their regulatory bodies. These usually are binding upon the produce industry as a matter of law and are used not only for the initiation of a business transaction but also for settlement of disputes if the transaction is not completed successfully. It is important to note that official standards still do not exist for every produce item. In these cases, disputes have to be resolved within the companies that conduct trade. Industry standards may be set by commodity producer groups who typically wish to set a standard for their product that may be higher than the official standard. These may be voluntary for producers, but those producers who do not participate in the program may not receive certain benefits that are offered to members of the group. If and when these standards are accepted widely by a commodity group they may be adopted as law and thus become official standards.

Association standards, as the term implies, are those set by trade associations for their members. Buyer standards, or specifications, are those set by businesses who wish to establish their own standards to generate customer confidence and loyalty. Although these specifications are not necessarily based on law, they have become a powerful tool in trade. In the area of food safety in particular, many retailers have imposed specific requirements upon their suppliers for certain conditions to be met. These usually involve audits or inspections of farms and facilities to assure compliance with the buyer's requirements, particularly food safety requirements. Consumer standards are not written or formal, but they may be the most important of all standards. These are the criteria that the consumers generally consider to make decision before purchasing a product.

Grading

Grading is a voluntary program of classification of a product based on certain quality characteristics like size, shape, colour, maturity, appearance *etc.* Types of grade of different commodities vary in different countries. This gives those in the produce industry common language for buying and selling. Grading may be mandatory for products that are subject to a marketing order, marketing agreement, or that are subject to import or export requirements. Grading is most commonly conducted at the packing or shipping point but may also be implemented by the receiver to settle disputes about quality. Grading also refers to the process conducted by company employees to remove inferior products prior to the manufacture of the package that will enter commerce. This is an essential step for quality assurance.

Fresh-cut Fruits and Vegetables: Quality and Safety

Fresh-cut fruits and vegetables are quickly growing part of the fresh produce market. Fresh-cut fruits and vegetables are freshly cut, cleaned, packed, and chilled produce. They are raw and, while being barely minimally processed, stay fresh and ready to eat or prepare. High levels of quality, together with increased security, are required for long-term industry growth and fresh-cut product consumption. The physiological, handling and storage needs of fresh-cut fruit and vegetables differ from those of typical, whole fruits and vegetables. In terms of physiology, handling, and storage, fresh-cut fruits and vegetables differ from traditional, intact fruit and vegetables. Due to peeling and cutting, fresh-cut fruits and vegetables are prone to both physiological and microbial damages. Browning may occur in fresh-cut produce when cell wall breaks down due to minimal processing, and phenolases, the enzymes present outside the cell wall and colourless phenols present inside the cell mix, causing a brown colour to occur. This is a chemical reaction that is referred to as enzymatic oxidative browning. As the name indicates, oxygen is a catalyst in the reaction. In addition to oxygen, presence of an enzyme, the polyphenol oxidase, copper, and a neutral pH hasten the reaction. The result of the chemical reaction is a substance with a brown pigment called as melanin. The damage of tissue and cell structure caused by fresh-cut treatment reduces the storage life of fruits and vegetables. As a result of the amplitude of enzymatic and respiration variables, and also microbial problems that influence safety, fresh-cut produce demands specific attention. Understanding the nature of fresh-cut produce as it relates to pre- and postharvest factors; handling, packaging, and storage is critical to achieving their quality, freshness, and nutritional value. Continuous efforts are to be made to develop the most appropriate techniques and technological innovations for sustaining their quality and ensuring safety to meet rising consumer demand.

Fresh-cut processing techniques and treatments have considerable influence on sensory quality including appearance, texture, taste and aroma of fruits and vegetables. Gil *et al.* (2006) recommended complete removal of the mango skin (peel) with a very sharp knife or peeler to avoid brown discoloration of the remaining peel tissues, which appears faster than flesh tissue browning of fresh-cut mango products. Chonhenchob *et al.* (2007) reported that extended shelf-life was observed in fresh-cut mangoes packed in PET due to reduced oxygen and elevated carbon dioxide concentrations. Quality of fresh-cut fruit and vegetable products determines the value to the consumer and is a combination of parameters including appearance, texture, flavour, and nutritional value (Kader, 2002). The colour, flavour, texture, and the nutritional value of fresh-cut fruit and vegetable products are factors critical to consumer acceptance and the success of these products (Barrett, *et al.*, 2010). Fresh cut products are at more risk of contamination and degradation due to application of various preparation steps such as washing, peeling, cutting, and slicing. Further, some processing methods aimed to avoid contamination and to extend the shelf-life of such products may sometimes act adversely, causing risk to human health (Yousuf *et al.*, 2020).

As the consumers have become more concerned about the nutritional and sensory aspects, as well as the safety of the food, the demand for healthy and nutritious food has been increased. In the affluent societies, the demand for convenience products including fresh-cut fruits and vegetables has also been increasing very rapidly. However, microbial contamination may occur in fresh-cut produce, which may cause serious hazards to human health, especially food-borne illness after consumption of contaminated fresh-cut fruits and vegetables. Thus, assurance of food safety is an important aspect which needs to be prioritized. As reviewed by Quadri *et al.* (2015), fresh-cut produce is more susceptible to spoilage and can facilitate rapid growth of spoilage microorganisms as well as the microorganisms of public health significance. Abadias *et al.* (2006) also opined that critical steps like proper procedures of production and disinfection were to be closely monitored to ensure the safety of fresh-cut fruits and vegetables.

Food Safety

Safety is a component of quality of food. It is better to mention that safety is the most important component of quality since a failure to assure safety can cause serious injury or even death to the consumer. It is difficult, usually impossible, to determine if a product is safe by simply looking at it from outside.

Unsafe food can affect everyone but the infants, young children, pregnant women, the elderly and those people with the underlying illness are particularly vulnerable posing global health threats. In any society, food safety needs to be treated as a public health priority for implementing effective food safety systems. All food handlers and consumers need to understand the roles they must play for food safety. Fruits and vegetables can become contaminated at any point of production, distribution and preparation. Though the primary responsibility lies with food producers, all food handlers and consumers also need to adopt basic hygienic practices when buying, selling and preparing food to protect their health as well as the health of the wider community.

Food safety is defined by the World Health Organization (WHO) as the assurance that the food will not cause harm to the consumer when it is prepared and/or eaten according to its intended use. The agents responsible for food borne diseases include bacteria, viruses, parasites, toxins, various chemical agents including pesticide residues and adulterants. So, quality of fruits and vegetables should be maintained properly to prevent the consumers from foodborne illnesses, commonly referred to food poisoning categorized under gastroenteritis, vomiting, diarrhoea *etc.* Reasonable assurance of safe food requires the reduction of risks that may occur during production, handling, and preparation for consumption by the end-user. Identifying risks is critical to food safety because it is difficult to control a hazard if it is not known to exist. The point of GAP and GMP is to reduce risks that occur during production and handling of fruits and vegetables in order to minimize the occurrence of illness or injury. Food safety cannot be achieved by responsibility of single industry or regulatory agency but the role of everyone is important as shown in Figure 3.1.

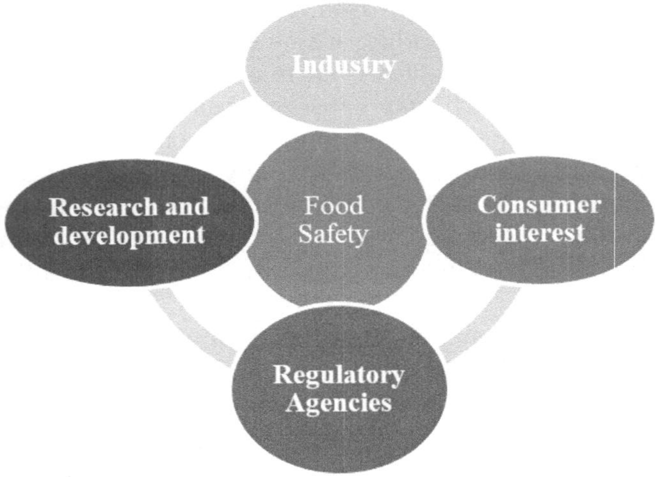

Figure 3.1: Inter-relationship of different Components of Food Safety.

Food safety has become a new concern in today's world due to awareness among the consumers, and active participation of various regulatory agencies (Boindala and Lewis, 2019). Different agents responsible for foodborne diseases are shown in Table 3.1.

for chemical hazards include effective, facility-specific good manufacturing practices (GMPs), food security and other prerequisite programmes; proper labelling and understanding of all components of ingredients and rigorous control of non-ingredient chemicals.

Chemical hazards in fresh produce can come from three general sources *viz.* naturally occurring substances, agricultural chemicals, and non-agricultural pollutants. Harmful chemicals have been associated with acute toxic responses and with chronic illnesses. There are many compounds in nature that cause harm to people if inhaled, ingested, or by contact with the skin, eyes or mucous membranes. Allergens can cause rapid and acute chemical toxicity. Mycotoxins *i.e.*, aflatoxin, mushroom toxins, phytohemoglutinin and some alkaloids are naturally occurring substances that can be toxic to people. The injury caused by some natural plant-associated toxins is triggered on skin by exposure to sunlight. People who are sensitive to any natural substance must take precautions to avoid exposure. In the case of known allergies, people should consider having suitable medical remedies available, such as injectable epinephrine, in the event of accidental exposure. Producers and handlers of fruits and vegetables should inspect fields and harvested products for any sign of contaminants with potential allergens or toxic agents. Agricultural chemicals of concern include pesticides, fertilizers and in animal production, antibiotics. Pesticides are the greatest concern for fruit and vegetable producers. Growers should always read and follow the instructions on pesticide labels. Handlers of pesticides should have suitable protective clothing and equipment and take care to use protective measures diligently.

Packing and packaging materials are potential sources of chemical hazards and these must not be allowed to enter foods. Plastics, vinyl chloride, paints and dyes, adhesives, lead and tin all are used in different types of packaging materials. These products should be manufactured in a manner that does not present hazards for consumers. Information regarding the potential for chemical transfer following exposure to heat, solvent vapours, oxidizing or reducing agents, or UV light should be provided by the manufacturers. Producers and handlers of fresh and fresh-cut fruits and vegetables should conduct a systematic and comprehensive risk assessment for their operations and develop procedures to minimize the potential exposure of consumers to hazards. The implementation of GAP and GMP can help to minimize disease occurrence. The grower must have a system for tracing product back to the field in which it was grown and tracing forward to the buyer or receiver.

Physical Hazards and Safety

Physical hazards in food processing can include such objects as metal fragments, nuts and bolts, glass and plastic, wood splinters, nut shells and fruit pits, and bone fragments. Some common control strategies for this category include -

- ☆ Physical exclusion of the hazard
- ☆ Effective detection and elimination systems
- ☆ Proper equipment design and selection
- ☆ Effective facility maintenance
- ☆ Visual inspection.

Once hazards have been identified, an effective programme can be designed to manage and reduce these risks. There are various Good Manufacturing Programmes such as HACCP, government regulations, FSSAI *etc.* which needs to be implemented strongly. Identification processes must be designed to recognize weaknesses in all of these areas.

Different possible hazards during processing of fruits and vegetables have been shown in Table 3.3.

Table 3.3: Different Possible Hazards in Processing

Processing Stage	*Physical Contamination*	*Chemical Contamination*	*Biological Contamination*
Receiving of fruit	Sand, pest, stones, wood	Naturally occurring substances Pesticides, fungicides Toxic chemicals, mycotoxin.	Presence of microorganism
Washing/Cleaning	Poorly maintained equipment	Chemical contaminations in water	Microbial quality of water Cross contamination Quality of water Cross contamination.
Peeling	Presence of metal pieces	Chemical contaminations from chemical or pesticides	Contamination due to unclean equipment
Pulping	Presence of metal pieces, foreign body or dust contamination from production environment.	None	Contamination due to un clean equipment.
Mixing	Screw, metal pieces	None	Cross contamination
Thermal processing	Unsanitized equipment	Sanitizer residue	Microbial survival due to insufficient temperature or holding time.
Storage	From warehouses	None	Cross contamination

Food Safety Strategies

Recent advances in science of health and food is somewhat effective in managing the various food-borne illness but at the same time the prevalence of disease is still there. In India, there is limited data available on the food-borne disease and its impact on public health. All these problems continue emphasizing the importance of ensuring food safety. The main components of a national food control system are:

- ✰ Implementation of food law and regulations
- ✰ Efficient management of food safety
- ✰ Inspection and identification of risk
- ✰ Updated laboratory services for quick monitoring and epidemiological data
- ✰ Information, education, communication and training to every participants of industry

Good Agricultural Practices (GAP) and Good Manufacturing Practices (GMP) in Maintaining Quality and Safety

Good Agricultural Practices (GAP) and Good Manufacturing Practices (GMP) are basic things in maintaining quality and safety of fresh and processed foods. All the workers engaged in production, handling and processing need to be trained on GAP and GMP to ensure that workers understand the importance of food safety and have the skills to utilize their knowledge.

Customers who purchase produce can expect a safe product. Marketing of adulterated product or presence of chemical, biological or physical contamination is illegal, and is regulated in many countries. Producers and Processors should feel a moral and ethical obligation to protect the health and safety of consumers. Training is one recognized step toward ensuring the sustainability of the company by exercising due diligence in food safety and minimizing the chance of being responsible for illness or outbreaks traced back to the operations. A work environment in which people are encouraged to share knowledge conveys the importance that all workers must share the responsibility for food safety. Training on GAP and GMP is an essential process to ensure that workers have a uniform foundation of skills for the safe production and handling of food.

A food safety hazard in any category among biological, chemical and physical hazards can cause harm to the consumer. Prevention of the occurrence of a hazard needs to be emphasised more than any type of remedial action to correct a problem after it has occurred. This is especially true for biological hazards *viz.* contamination with microorganisms because there is no provision for complete inactivation of all microbes that can be present on fresh produce. The key to prevention is education and effective training followed by implementation of GAP and GMP.

Safety and Quality Assurance

Fresh fruit and vegetable growers, packers, shippers, retailers and consumers all have long recognized, at least intuitively, that product quality is the primary factor affecting profitable trade in the produce industry. Only in the past twenty years or so has food safety also become a driving force in conducting business. This has resulted in a plethora of opinions regarding the relationship of safety to quality. In commerce perhaps the safety factor is not often given greater importance than the overall quality, but when an outbreak of illness occurs, safety overrides all other quality considerations.

Nutritional value is related to the presence and amounts of constituents that support life. Fresh fruits and vegetables are recognized as good sources of vitamins, minerals and fiber. More recently researchers have identified that they are sources of antioxidants and other phytochemicals that have roles in preventing or controlling some diseases. The types, quality and quantity of these constituents that individuals consume can directly impact their health. Safety, defined earlier, is the assurance that food will not cause harm to the consumer when it is prepared and/or eaten according to its intended use. Detecting and monitoring safety risks are difficult. Microbiological techniques are used to determine the presence of pathogens. Chemical analysis is required to detect pesticides or other chemicals. Physical hazards may be found by x-ray or other imaging techniques. Some tests are destructive so not every piece of product can be tested. The implementation of an effective safety assurance program is essential in order for a test to reasonably predict the safety of the entire lot of product.

International Standards

There are many regulating bodies that have set standards for international trade. International standards and different codes for various crops are developed by different international organizations to ensure quality and safety of food.

International organisations working in ensuring food safety are:

1. World Trade Organization (WTO)
2. International Standards Organization (ISO)
3. Codex Alimentarius Commission (Codex)
4. Office International des Epizooties (OIE)
5. International Plant Protection Convention (IPPC)

In conjunction with 'World Trade Organization (WTO) Agreements', the 'Codex Committee on Fresh Fruits and Vegetables' has the responsibility for developing standards and codes of practice for fresh produce. A code of practice known as "Quality Inspection and Certification of Fresh Fruits and Vegetables" has been adopted by the Codex Alimentarius Commission.

This code contains provisions for packing, shipping, control and inspection of fresh produce. Codex standards are a combination of grading for quality and inspection for wholesomeness, safety and freedom from economic fraud. Inspection and certification are conducted at the point of origin by a national official or a recognized service person. The objective of the Codex standards is to protect consumers' health and ensure fair practices in the trade of food. The Codex Committee on Food Import and Export Inspections recommends that public health protection issues be given the highest priority when considering standards. There is no legal obligation for World Trade Organization (WTO) members to adopt Codex Standards into law. Individual countries establish their own standards for imported food. However, member countries must be prepared to justify to WTO any domestic standard that is more restrictive to trade than the Codex standard.

The International Organization for Standardization (ISO) is another well-known entity in many industries including food. ISO specifies requirements for food safety management systems and requires that they demonstrate the ability to control food safety hazards in order to ensure that food is safe at the time of consumption. ISO standards are applicable to all organizations regardless of size. Phytosanitary issues are a concern for all importing countries. Domestic agencies are responsible for protecting their own food supplies from the import of exotic pests that may threaten domestic production.

World Health Organization (WHO) Response

World Health Organization (WHO) aims to facilitate global prevention, detection and response to public health threats associated with unsafe food. WHO works to ensure consumer trust in their authorities, and confidence in the safe food supply.

To do this, WHO helps Member States build capacity to prevent, detect and manage foodborne risks by:

- ✰ Providing independent scientific assessments on microbiological and chemical hazards that form the basis for international food standards, guidelines and recommendations, known as the Codex Alimentarius, to ensure food is safe wherever it originates;
- ✰ Assessing, in structured, transparent and measurable ways, the performance of food control systems throughout the entire food chain, identifying priority areas for capacity development, and measuring and evaluating progress over time through the FAO/WHO food control system assessment tool;
- ✰ Assessing the safety of new technologies used in food production, such as genetic modification and nanotechnology;
- ✰ Helping improve national food systems and legal frameworks, and implement adequate infrastructure to manage food safety risks. The

International Food Safety Authorities Network (INFOSAN) was developed by WHO and the UN Food and Agriculture Organization (FAO) to rapidly share information during food safety emergencies;

- Promoting safe food handling through systematic disease prevention and awareness programmes, through the WHO Five Keys to Safer Food message and training materials; and
- Advocating for food safety as an important component of health security and for integrating food safety into national policies and programmes in line with the International Health Regulations (IHR) - 2005.

World Health Organization works closely with the Food and Agriculture Organization (FAO), the World Organization for Animal Health (OIE) and other international organizations to ensure food safety along the entire food chain from production to consumption.

Globalization has triggered growing consumer demand for a wider variety of foods, resulting in an increasingly complex and longer global food chain. As the world's population is growing with a rapid pace, the intensification and industrialization of agriculture and animal production to meet increasing demand for food creates both opportunities and challenges for food safety. Climate change is also predicted to impact food safety in future. These challenges put greater responsibility on food producers and handlers to ensure food safety.

References

Abadias, M., Canamas, T. P., Asensio, A., Anguera, M. and Vinas, I. 2006. Microbial quality of commerecial 'Golden Delicious' apples throughout production shelf-lifein Lleida (Catalonia, Spain). *Int. J. Food Microbio.,* 108: 404-409.

Barrette, D. M., Beaulieu, J. C. and Shewfelt, R. 2010. Color, flavor, texture, and nutritional quality of fresh-cut fruits and vegetables: desirable levels, instrumental and sensory measurement, and the effects of processing. Crit Rev Food Sci Nutr, **50** (5): 369-389.

Boindala, S. and Lewis, J. 2019. The grand challenge of regulating health foods in India. Indian *J. Med. Res.,* **150** (3): 248-253.

Chonhenchob, V., Chantarasomboon, Y. and Singh, S. P. 2007. Quality changes of treated freshcut tropical fruits in rigid modified atmosphere packaging containers. *Packag. Technol. Sci.,* 20: 27-37.

Gabriel OA, Puleng L (2013). Strategies for the Prevention and Reduction of Mycotoxins in Developing Countries. Mycotoxin and Food Safety in Developing Countries.

Gil, M.I., Aguayo, E. and Kader, A. A. 2006. Quality changes and nutrient retention in fresh-cut versus whole fruits during storage. *J. Agric. Food Chem.* 54: 4284-4296.

Kader, A.A. 2002. Quality parameters of fresh-cut fruit and vegetable products. In: *Fresh-cut Fruits and Vegetables: Science, Technology, and Market*, O. Lamikanra (Ed.), CRC Press, Boca Ratón, USA. pp. 11-20.

Qadri, O. S., Yousuf, B. and Srivastava, A. K. 2015. Fresh-cut fruits and vegetables: Critical factors influencing microbiology and novel approaches to prevent microbial risks—A review. Cognet Food and Agriculture, 1(1): 1-17.

Mitra, S. 2021. Postharvest technology-An overview. **In**: *Postharvest Management of Horticultural Crops.* Surajit Mitra (Ed.), Jaya Publishing House, New Delhi-110089, India, pp. 1-16.

Mitra, S., Mitra, S. and Tarafdar, J. (2021). Antioxidant substances and phytonutrients in sweet potato tubers of different flesh colour. *Ann. Phytomed.*, **10**(2): 1-7. http: //dx.doi.org/10.21276/ap.2021.10.2.

Paull, R.E. and Chen, N.J. 2000. Heat treatment and fruit ripening. *Postharvest Biol. Technol.*, 21: 21-37.

Pérez, A.G., Luaces, P., Ríos, J.J., García, J.M. and Sanz, C. 2003. Modification of 434 volatile compound profile of virgin olive oil due to hot-water treatment of olive fruit. *J. Agric. Food Chem.*, **51**: 6544-6549.

Yousuf, B., Deshi, V., and Siddiqui, M. W. 2020. Fresh-cut fruits and vegetables: Quality issues and safety concerns. In: *Fresh-cut Fruits and Vegetables-Technologies and Mechanisms for Safety Control,* Siddiqui, M. W. (Ed.), Academic Press, pp.-1-15.

Websites

https: //www.fao.org/food-safety/en/

https: //www.who.int/health-topics/food-safety

Chapter 4

Pre-harvest and Postharvest Factors Influencing Quality and Shelf Life of Fruits and Vegetables

Introduction

Pre-harvest factors have tremendous impact on quality attributes of fruits and vegetables both at harvest and postharvest life. Whereas, the postharvest factors play an important role in maintaining the quality rather than improving it. Among the several factors responsible for the overall quality of produce, about 70 per cent comes from pre-harvest conditions while only 30 per cent of postharvest factors affect the quality. Quality losses during postharvest life are results of various pre-harvest factors as well as postharvest factors. Therefore, it is very important to give proper attention to different pre-harvest factors that influence the important harvest quality attributes, postharvest quality and the marketability of the harvested produce. The production management practices including genetic factors, plant nutrition, irrigation, plant protection *etc.* should be given proper importance for production of high quality fruit and vegetables.

Tropical fruits and vegetables generally harvested at relatively high temperatures containing large amounts of field heat are metabolically active and undergo rapid ripening and senescence processes during postharvest stages. Inadequate management of these processes can result in major losses in nutritional and quality attributes and outbreaks of foodborne pathogens causing huge postharvest losses. Proper management of both pre-harvest and postharvest factors are crucial to slow down physiological processes and to minimize the risk of microbial growth and contamination, which in addition, help in maintaining postharvest quality and prolonging shelf life the produce. Providing optimum conditions for cropping, time and stage of harvest, postharvest handling, storage conditions and marketing methods are also important in determining final product quality at the consumer level.

Pre-harvest Factors

Pre-harvest factors *viz.*, genetic factors, environmental factors like temperature, relative humidity, water potential, light, and cultural practices and pest management techniques determine the inherent quality of the produce. However, the ultimate quality is the final manifestation of inter relation between the commodity and its environment. Some of these factors are related to plant, others are related to environment or to cultural practices.

1. Genetic Factors

Types of Crop

Quality of the fruits and vegetables varies from crop to crop due to the differences in inherent quality, *e.g.* in potato, onion, pumpkin and garlic, good quality is related to shelf life. While apple, mango, cherry, strawberry, tomato, capsicum *etc.* are attractive to consumers due to their external appearance, colour, taste and flavour. Genetic factors underlying drought tolerance in *Solanum chilense* (Dunal) Reiche and *S. pennellii* Correll, two wild tomato species indigenous to arid and semiarid environments of South America, are being explored by International Crops Research Institute for the Semiarid Tropics (ICRISAT) and Asian Vegetable Research and Development Centre (AVRDC) with the aim of transferring drought tolerance into cultivated tomatoes. AVRDC has made significant contributions to the development of heat-tolerant tomato and Chinese cabbage (*Brassica rapa* L.) lines and the subsequent release of adapted, tropical varieties worldwide. However, lower yields in heat-tolerant lines remain a concern. Cultivation of indigenous vegetables, fruits, and other underutilized crops is being promoted as many of these plants naturally have higher levels of resilience to environmental stress. These underutilized crops can grow and produce adequately under poor soil conditions, for example in drought-prone or saline areas, and often are more resistant to pests and diseases. In many instances, indigenous vegetables can grow in environments unfit for other crops. They are very strong candidates to withstand the adverse effects of a changing climate.

Cultivars

The quality of seed or planting material is an important factor that controls the postharvest quality of the fruits and vegetables, some of which are controlled genetically. Growers need to select the cultivars as per the desired quality and intended use prior to planting of crops. This choice may be limited by availability of planting material depending on the crop. Preference of the consumers and demand in the market should also be kept in mind prior to selecting a cultivar of any crop. Postharvest product quality develops during growing of the product and is maintained, not improved by postharvest technologies. Available genetic material allows discrimination of external and internal quality attributes that must satisfy consumer requirements. In some crops, especially in commercial fruit and vegetable crops, wide range of varieties with different quality attributes are available to growers for planting. If available, a variety with good shape, size, colour, taste, productivity, dry matter, and postharvest longevity can be selected for planting. Commercial varieties having good quality attributes but short shelf life can be used for combining qualitative traits in the varieties of very long shelf life through breeding and genetic engineering.

Nevertheless, rootstocks are decisive not only for the vigour and yield of the grafted cultivar (Avdiu *et al.*, 2014; Domi *et al.*, 2014; Spahiu *et al.*, 2013a, b; Thomaj *et al.*, 2015), but also size, acidity, color, and firmness of the fruit (López-Ortega *et al.* (2016)). Grafting of watermelon is an effective tool to overcome continuous cropping obstacles and control soilborne diseases (Lee, 1994; Miguel *et al.*, 2004; Paroussi *et al.*, 2007; Rouphael *et al.*, 2010; Yetisir *et al.*, 2003). Another grafted vegetable, eggplant, has a series of rootstocks, each conferring resistance to various diseases and other resistances. *Solanum melongena* lines and hybrids resist *Fusarium* and bacterial wilt (Monma *et al.*, 1997; Yoshida *et al.*, 2004a, b), controls *Phomopsis* blight, confers windfall resistance (MeiXiu *et al.*, 2001), as well as variation of fruit epidermis and flesh firmness (Suzuki *et al.*, 2004). *Solanum integrifolium* resists *Fusarium* and bacterial wilt (Mochizuki and Yamakawa, 1979a; Yoshida *et al.*, 2004a).

2. Environmental Factors

Along with many pre-harvest factors, environmental factors like temperature, rains, relative humidity, water potential, and light intensity and duration also determine the inherent quality of the fresh produce. As per the adaptation of the plants to grow in a particular climate and the zones where maximum productivity is achieved, they are classified as tropical, subtropical, warm temperate and cool temperate crops. The most crops have a temperature optimum in which they attain maximum yields and quality. If these crops are planted outside fairly specific temperature and day length criteria, limited growth occurs. Tropical or subtropical crops suffer from chilling or freezing injury if grown in temperate climates. Also, temperate crops will not flower successfully and may sustain heat injury if grown in tropical climates.

Environmental stress is a primary cause of crop losses, reducing average yields for most major fruit and vegetable crops by more than 50 per cent (Boyer, 1982; Bray *et al.*, 2000). It also affects the quality of fresh produce considerably. The severity of environmental stress imposed on crops is influenced by climatic uncertainty and change. Extreme geophysical events such as more erratic rainfall patterns, unpredictable high temperature spells, reduced water availability, climate change due to excess emission of greenhouse gases may also be responsible for substantial loss in production and quality of produce. Adapting agricultural management practices to the production challenges posed by climate variability and change such as limited rainfall and/or irrigation water, flooding, and salinity is vital for future harvests. The International Centre for Agricultural Research in the Dry Areas (ICARDA) promotes protected agriculture in arid regions with controlled growth environments using less water and fertilizer and careful production timing to improve yield. AVRDC has also developed protected production systems for use with excessive rainfall and grafting to increase flood tolerance. Management strategies have also been developed such as modification of fertilizer application, efficient delivery of water to roots through water-saving irrigation methods, and use of soil amendments to enhance nutrient uptake by plants and improve soil fertility.

Taste and flavour of horticultural crops are influenced by the environment, agrichemicals, nutrition and management systems that can impact on flavour through effects on plant development (Mattheis and Fellman, 1999). Appropriate light (intensity and quality) and temperature do influence postharvest eating quality (Kays, 1999). Both are required for optimal plant productivity and harvest index (the edible dry weight – dry matter or DM – of the crop harvested as a proportion of the total dry weight of the plant). In perennial crops, light utilisation is a key determinant of productivity and quality (Snelgar *et al.*, 1998; Tustin *et al.*, 2001) as the leaf area exposed to sun during the day must be adequate to provide the carbon needed for both fruit and vegetative growth. However, the radiation, temperature and possibly rainfall conditions during the growing season are such that maximal partitioning of sugars (and other components that contribute to the important quality attributes of wine) are accumulated in the fruit, instead of the shoots, prior to harvest. There is a real need to develop more robust physiological models of carbon assimilation, distribution and accumulation on different plant organs of commercially important crops. The effects of environmental factors on quality of different fruits and vegetables are given in Table 4.1.

3. Cultural Practices

Pre-harvest production practices may considerably affect postharvest returns in terms of quality and quantity. All the cultural practices have direct effect on the final quality of the produce.

Planting Period

Many plants are very sensitive to environmental conditions, and thus quality will not be optimized when crop is produced under adverse conditions. Producing summer plants during the winter or winter crops during summer will not be practicable or profitable, unless grown under protection. Protected cultivation is often used as a means of amending unfavourable growing

Table 4.1: Environmental Factors on Quality of Fresh Produce

Sl.No.	*Environmental Factors*	*Quality Affected*
1	Temperature	High temperature affects maturity, colour, sugar, acidity *etc.*; reduce the quality *e.g.* in citrus, radish, spinach cauliflower *etc.* and increase the quality in grapes, melons tomato *etc.* Low temperature cause chilling and freezing injury which reduce the quality.
2	Light	Essential for anthocyanin formation. Exposed fruit to sun light develop lighter weight, thinner peel, lower juice and acids and higher TSS than shaded fruits, *e.g.* citrus mango *etc.*, exposure of potato to light causes greening (solanine formation) which has toxic properties. High sun light intensity causes sun scaled in citrus and tomato and reduce the poor white colour in cauliflower. Low light intensity causes thin and large leaves in leafy vegetables.
3	Rains	Rains after a dry spell causes cracking in grapes, dates, litchi, limes, lemon, tomato, sweet potato *etc.* It reduces appearance and sweetness.
4	Winds	Causes brushing, scratching and corky scar (citrus fruits), and damage leafy vegetables.
5	Humidity	High humidity reduces the colour and TSS and increases acidity in citrus, grapes, tomato, *etc.* but it is needed for better quality of banana, litchi, and pineapple.

Source: Srivastava, 2002.

conditions to produce crops that are generally destined for specific markets as 'off season' crops when the same crops grown in open fields are not available. A diverse range of horticultural crops are grown under glass or plastic covers. In general, the physiology of all these crops grown under protective covers is the same as that of similar crops grown outdoors, and the final quality is influenced by pre-harvest conditions. The quality deterioration begins after harvest and is markedly influenced by postharvest temperature, relative humidity, ethylene and atmosphere. These crops are grown in a sheltered environment and not exposed to variable wind, rain and outside temperature conditions. As the produce grown under cover is often more susceptible to physical damage during postharvest handling than the produce grown in the open field, more care should be taken during harvesting and postharvest handling to avoid mechanical injury.

Planting Density

It has direct effect on both the quantity and quality of the produce. High density planting increases competition between plants, reduces light availability, and thus may decrease both quantity and quality. Low density planting lead to large size, better colored fruit or vegetable which may have shorter shelf life as larger fruits are commonly more sensitive to physiological disorders. Tree form will influence the amount of light perceived and captured by fruit trees. Traditional horticultural technologies, using size controlling rootstocks and adaptive pruning and thinning techniques, are good examples of enabling technologies to optimise productivity (crop load) and quality of fruit. These technologies do so by influencing size and form of the tree and hence the volume occupied in an orchard. Conceptual advances in understanding the genetic plasticity of trees, or the extent to which the external or internal environment can influence tree architecture, should provide further insights in factors that control carbon assimilation and distribution within fruit trees (Seleznyova *et al.*, 2003).

Water Supply/Irrigation

Growing plants need a continuous water supply for both photosynthesis and transpiration. Both too much rain or irrigation and lack of rain or irrigation are detrimental to plants for their growth, yield and quality. Too much rain or irrigation can lead to brittle and easily damaged leafy vegetables with an increased tendency to decay. Whereas, lack of rain or irrigation can lead to low juice content and thick skin in some fruits, *e.g.* citrus. Long dry conditions followed by rain or irrigation may cause growth cracks or secondary growth in potatoes, cracks in tomatoes, litchi fruits *etc.* Irregular watering usually reduces fruit size, increases splitting and physiological disorders, and reduces water content in the plant or plant parts.

In general, growers adopt water management strategies to minimise moisture stress so as to allow optimal photosynthesis, plant growth and crop yield. Irrigation systems vary, but in all cases, the good quality water should be used. Water needs to be applied on a regular basis determined by the evapotranspiration demand well before serious stress conditions occur. In some situations with some crops, careful manipulation of water supply may decrease water usage and improve crop quality without compromising sustainable plant growth. However, maximum yields are not always a prerequisite for optimal quality. Having very high yields in grapes can compromise wine quality. The best wine quality vintages come from those years where environmental conditions often impose stress on the vines (high temperatures, low rainfall). The use of Regulated Deficit Irrigation (RDI) to minimise water applied without affecting plant performance, and sometimes increasing fruit quality, is a system that has been used successfully for a range of fruit crops (Behboudian and Mills, 1997).

Soil Fertility/Plant Nutrients

Lack of plant nutrients in the soil can seriously affect the quality of fresh produce at harvest. On the other hand, application of excess fertilizers can harm the development and postharvest condition of fresh produce. Lack of nitrogen in the soil can lead to stunted growth or to the yellow-red discoloration of leaves in green vegetables, *e.g.* cabbage. Lack of potash may cause poor development and abnormal ripening in many fruits. Calcium-moisture imbalance can cause